Springer Complexity

Springer Complexity is an interdisciplinary program publishing the best research and academic-level teaching on both fundamental and applied aspects of complex systems—cutting across all traditional disciplines of the natural and life sciences, engineering, economics, medicine, neuroscience, social and computer science.

Complex Systems are systems that comprise many interacting parts with the ability to generate a new quality of macroscopic collective behavior the manifestations of which are the spontaneous formation of distinctive temporal, spatial or functional structures. Models of such systems can be successfully mapped onto quite diverse "real-life" situations like the climate, the coherent emission of light from lasers, chemical reaction-diffusion systems, biological cellular networks, the dynamics of stock markets and of the internet, earthquake statistics and prediction, freeway traffic, the human brain, or the formation of opinions in social systems, to name just some of the popular applications.

Although their scope and methodologies overlap somewhat, one can distinguish the following main concepts and tools: self-organization, nonlinear dynamics, synergetics, turbulence, dynamical systems, catastrophes, instabilities, stochastic processes, chaos, graphs and networks, cellular automata, adaptive systems, genetic algorithms and computational intelligence.

The three major book publication platforms of the Springer Complexity program are the monograph series "Understanding Complex Systems" focusing on the various applications of complexity, the "Springer Series in Synergetics", which is devoted to the quantitative theoretical and methodological foundations, and the "Springer Briefs in Complexity" which are concise and topical working reports, case studies, surveys, essays and lecture notes of relevance to the field. In addition to the books in these two core series, the program also incorporates individual titles ranging from textbooks to major reference works.

Series Editors

Henry D. I. Abarbanel, Institute for Nonlinear Science, University of California, San Diego, La Jolla, CA, USA
Dan Braha, New England Complex Systems Institute, University of Massachusetts, Dartmouth, USA
Péter Érdi, Center for Complex Systems Studies, Kalamazoo College, USA and Hungarian Academy of Sciences, Budapest, Hungary
Karl J. Friston, Institute of Cognitive Neuroscience, University College London, London, UK
Hermann Haken, Center of Synergetics, University of Stuttgart, Stuttgart, Germany
Viktor Jirsa, Centre National de la Recherche Scientifique (CNRS), Université de la Méditerranée, Marseille, France
Janusz Kacprzyk, Polish Academy of Sciences, Systems Research Institute, Warsaw, Poland
Kunihiko Kaneko, Research Center for Complex Systems Biology, The University of Tokyo, Tokyo, Japan
Scott Kelso, Center for Complex Systems and Brain Sciences, Florida Atlantic University, Boca Raton, USA
Markus Kirkilionis, Mathematics Institute and Centre for Complex Systems, University of Warwick, Coventry, UK
Jürgen Kurths, Nonlinear Dynamics Group, University of Potsdam, Potsdam, Germany
Ronaldo Menezes, Department of Computer Science, University of Exeter, UK
Andrzej Nowak, Department of Psychology, Warsaw University, Warszawa, Poland
Hassan Qudrat-Ullah, School of Administrative Studies, York University, Toronto, Canada
Linda Reichl, Center for Complex Quantum Systems, University of Texas, Austin, USA
Peter Schuster, Theoretical Chemistry and Structural Biology, University of Vienna, Vienna, Austria
Frank Schweitzer, System Design, ETH Zürich, Zürich, Switzerland
Didier Sornette, Entrepreneurial Risk, ETH Zürich, Zürich, Switzerland
Stefan Thurner, Section for Science of Complex Systems, Medical University of Vienna, Vienna, Austria

Understanding Complex Systems

Founding Editor: S. Kelso

Future scientific and technological developments in many fields will necessarily depend upon coming to grips with complex systems. Such systems are complex in both their composition–typically many different kinds of components interacting simultaneously and nonlinearly with each other and their environments on multiple levels–and in the rich diversity of behavior of which they are capable.

The Springer Series in Understanding Complex Systems series (UCS) promotes new strategies and paradigms for understanding and realizing applications of complex systems research in a wide variety of fields and endeavors. UCS is explicitly transdisciplinary. It has three main goals: First, to elaborate the concepts, methods and tools of complex systems at all levels of description and in all scientific fields, especially newly emerging areas within the life, social, behavioral, economic, neuro- and cognitive sciences (and derivatives thereof); second, to encourage novel applications of these ideas in various fields of engineering and computation such as robotics, nano-technology, and informatics; third, to provide a single forum within which commonalities and differences in the workings of complex systems may be discerned, hence leading to deeper insight and understanding.

UCS will publish monographs, lecture notes, and selected edited contributions aimed at communicating new findings to a large multidisciplinary audience.

More information about this series at http://www.springer.com/series/5394

Iddo Eliazar

Power Laws

A Statistical Trek

 Springer

Iddo Eliazar
Tel Aviv, Israel

ISSN 1860-0832   ISSN 1860-0840 (electronic)
Understanding Complex Systems
ISBN 978-3-030-33234-1   ISBN 978-3-030-33235-8 (eBook)
https://doi.org/10.1007/978-3-030-33235-8

© Springer Nature Switzerland AG 2020
This work is subject to copyright. All rights are reserved by the Publisher, whether the whole or part of the material is concerned, specifically the rights of translation, reprinting, reuse of illustrations, recitation, broadcasting, reproduction on microfilms or in any other physical way, and transmission or information storage and retrieval, electronic adaptation, computer software, or by similar or dissimilar methodology now known or hereafter developed.
The use of general descriptive names, registered names, trademarks, service marks, etc. in this publication does not imply, even in the absence of a specific statement, that such names are exempt from the relevant protective laws and regulations and therefore free for general use.
The publisher, the authors and the editors are safe to assume that the advice and information in this book are believed to be true and accurate at the date of publication. Neither the publisher nor the authors or the editors give a warranty, expressed or implied, with respect to the material contained herein or for any errors or omissions that may have been made. The publisher remains neutral with regard to jurisdictional claims in published maps and institutional affiliations.

This Springer imprint is published by the registered company Springer Nature Switzerland AG
The registered company address is: Gewerbestrasse 11, 6330 Cham, Switzerland

ספר זה מוקדש להורי היקרים
לאבי אלי , בן מלכה ויצחק לבית אליעזר , זכרו לברכה
ולאימי שרה, בת רחל וברוך לבית ארונוב, תבדל לחיים ארוכים

עדו אליעזר
ערב ראש השנה התש"פ
ישראל

This book is dedicated to my dear parents
To my father Eli, son of Malka and Isaac
Eliazar, may his memory be blessed
To my mother Sara, daughter of Rachel and
Baruch Aronov, may she be granted long life

Iddo Eliazar
Eve of the Jewish year 5780
Israel

Preface

In science and engineering in general, and in complex systems in particular, one encounters an assortment of *power statistics*: statistical representations, of empirical data, with inherent power-law forms. Examples of power statistics include: power distribution functions—characterizing the Pareto and inverse-Pareto laws; power rank distributions—characterizing the Zipf and inverse-Zipf laws; power hazard rates—characterizing the Weibull and Fréchet (inverse-Weibull) extreme-value laws; power Lorenz curves—manifesting socioeconomic fractality; power log-Laplace and log-Fourier transforms—characterizing the one-sided and symmetric Lévy laws; power first-digit distributions—generalizing the Newcomb–Benford law; power temporal mean-square-displacements of random motions—manifesting sub-diffusion and super-diffusion; power spectral densities of stationary random processes—termed flicker and 1/f noises. Remarkably, this assortment of power statistics stands on a common foundational bedrock: the *power Poisson process*. Namely, this bedrock is a Poisson process that is defined over the positive half-line, and that is governed by a power intensity function. Arguably, the power Poisson process is the "mother of all power statistics". Figuratively, the power Poisson process is an infinite iceberg, and the aforementioned power statistics are some of its tip-of-the-iceberg facets. This monograph presents a comprehensive and in-depth exploration of the power Poisson process: its host of statistical properties, and its emergence via stochastic limit laws. Also, constructed bottom-up from the power Poisson process, this monograph presents a unified and overarching exposition of power statistics. The investigations of the power Poisson process and of power statistics, presented herein, employ diverse perspectives: structural, probabilistic, fractal, dynamical, socioeconomic, and stochastic. This monograph is poised to serve researchers and practitioners—from various fields of science and engineering—that are engaged in power-statistics analyses.

Tel Aviv, Israel   Iddo Eliazar

Acknowledgements

This monograph has been stewing for about 20 years, during a long and fascinating journey of scientific exploration. Along this journey, I got to know and meet remarkable colleagues. Some of these colleagues became coauthors, some became close friends, and some both. I wish to acknowledge: Elena Agliari; Eitan Altman; Galit Ashkenazi Golan; Rami Atar; Amir Averbuch; Ido Bachelet, wizard of biomimetics; Olivier Benichou; Eric Berger; Onno Boxma, KNIGHT of the Order of the Netherlands Lion; Micha Breakstone; David Campbell—who gave me the final nudge to write this monograph; Sandra Chapman; Aleksei Chechkin; Pasquale Cirillo; Morrel Cohen, GRANDMASTER of theoretical physics; the late Maxime Dahan, who passed away so hastily and unexpectedly; Eli David; Israel David; Gadi Fibich; Andrea Fontanari; Giovanni Giorgi; Yoav Git; Rony Granek; Paolo Grigolini; Ted Groves; Ewa Gudowska Nowak; Shlomo Havlin; Sara Hershkovitz; Yurij Holovatch; Bruce Horn; Marie Jardat; Tal Kachman; Eugene Kanzieper; Ed Kaplan HAGADOL; Offer Kella; Ron Kenett; Yoav Kerner; Janos Kertesz; Sascha Khakshouri; Yossi Klafter—who lured me into the magical realm of anomalous phenomena; Rainer Klages; Tal Koren; Alex Ely Kossovsky; Lasse Leskela; Hanoch Levy; Varda Liberman; Uri Liberman; Katja Lindenberg, QUEEN of the anomalous realm; Abraham Lioui; Erez Livneh; Marcin Magdziarz; Avishai Mandelbaum; Oded Margalit; Gustavo Martinez Mekler; Mark Meerschaert; Rob van der Mei; Yasmine Meroz; Ralf Metzler—keep them sandals on; Tal Mofkadi; Carlos Mejia Monasterio—Panna Cotta; Ron Milo; Olesya Mryglod; Yoni Nazarathy; Gleb Oshanin; Arnab Pal; Karol Penson; Itamar Procaccia; Rami Pugatch; Sid Redner; Ben Reis; Shlomi Reuveni—whom I had the honor of co-supervising toward a Ph.D. in mathematics; Tamie Ronen; Sheldon Ross; the late Reuven Rubinstein; Shlomi Rubinstein; Eitan Sayag; Enrico Scalas; the late Zeev Schuss—who time and again advised me to write a monograph; Adam Schwartz; Roy Shalem; Nahum Shimkin; Mike Shlesinger, MASTER JEDI of the anomalous realm; Igor Sokolov—the one and only; Didier Sornette; Gene Stanley, founding father of Econophysics; Nassim Taleb; Alessandro Taloni; Charles Tapiero; Oren Tapiero; Tamir Tassa; Malvin Teich; Michael Urbakh; Maria

Vlasiou; Raphael Voituriez; Nick Watkins; Steve Weiner; Gideon Weiss; Alex Weron; Karina Weron; Horacio Wio; Liat Yakir; Uri Yechiali—my lifelong MENTOR and role model; Shlomo Yitzhaki; Ofer Zeitouni; and Bert Zwart. I also wish to acknowledge three educators who, many many years ago, sowed the seeds: **SARA ARONOV ELIAZAR**, my beloved Wonder-Woman mother; Thomas Goodman; and Uri Gothelf.

Contents

1	**Introduction**	1
	References	7
2	**From Lognormal to Power**	13
	2.1 Geometric Evolution	13
	2.2 Gaussian White Noise	14
	2.3 Brownian Motions	14
	2.4 Gaussian Motions	15
	2.5 Gaussian Mean and Variance	16
	2.6 Gaussian Stationary Velocities	17
	2.7 Poisson-Process Limit	18
	2.8 Outlook	19
	2.9 Notes	20
	2.10 Methods	22
	2.10.1 Equation (2.8)	22
	2.10.2 Equation (2.13)	23
	2.10.3 A General Poisson-Process Limit-Law	24
	2.10.4 The Power Poisson-Process Limit-Law	26
	References	27
3	**Setting the Stage**	31
	3.1 The Poisson Law	31
	3.2 Framework	32
	3.3 Methods	34
	3.3.1 Normal Limit for Poisson Random Variables	34
	3.3.2 Chernoff Bounds for Poisson Random Variables	35
	3.3.3 Reciprocation of $\mathcal{E}_+$ and $\mathcal{E}_-$	37
	References	37

4 Threshold Analysis ... 39
- 4.1 Mean Behavior ... 39
- 4.2 Asymptotic Behavior ... 40
- 4.3 Log-Log Behavior ... 41
- 4.4 Pareto Laws ... 42
- 4.5 Pareto Scaling ... 43
- 4.6 Truncated Weibull Laws ... 45
- 4.7 Weibull Laws ... 45
- 4.8 Outlook ... 47
- References ... 48

5 Hazard Rates ... 49
- References ... 51

6 Lindy's Law ... 53
- References ... 60

7 Order Statistics ... 61
- 7.1 Simulation ... 61
- 7.2 Statistics ... 62
- 7.3 Asymptotic Behavior ... 64
- 7.4 Zipf Laws ... 65
- 7.5 Ratios ... 66
- 7.6 Log-Ratios ... 67
- 7.7 Forward Motion ... 68
- 7.8 Backward Motion ... 69
- 7.9 Outlook ... 70
- 7.10 Methods ... 71
  - 7.10.1 Simulation ... 71
  - 7.10.2 Equations (7.3) and (7.4) ... 72
  - 7.10.3 Equations (7.9) and (7.11) ... 73
  - 7.10.4 Equations (7.12) and (7.14) ... 76
  - 7.10.5 Equations (7.20) and (7.21) ... 77
  - 7.10.6 Equations (7.22) and (7.24) ... 78
- References ... 79

8 Exponent Estimation ... 81
- References ... 86

9 Socioeconomic Analysis ... 87
- 9.1 Socioeconomic Perspective ... 87
- 9.2 Disparity Curve ... 88
- 9.3 Disparity-Curve Analysis ... 89
- 9.4 Lorenz Curves ... 90

	9.5	Lorenz-Curves Analysis	92
	9.6	Inequality Indices	93
	9.7	Gini Index	94
	9.8	Reciprocation Index	96
	9.9	Summary	97
	9.10	Methods	98
		9.10.1 Equation (9.4)	98
		9.10.2 Equation (9.5)	99
	References	99
10	**Fractality**	101
	10.1	Scale Invariance	101
	10.2	Perturbation Invariance	102
	10.3	Symmetric Perturbations	103
	10.4	Socioeconomic Invariance	104
	10.5	Poor Fractality and Rich Fractality	106
	10.6	Renormalization	107
	10.7	Summary	108
	10.8	Methods	109
		10.8.1 Equations (10.1), (10.2), and (10.11)	109
		10.8.2 Equations (10.3) and (10.4)	110
		10.8.3 Equations (10.5) and (10.6)	111
		10.8.4 Equations (10.7) and (10.8)	112
	References	113
11	**Sums**	115
	11.1	One-Sided Lévy Law I	115
	11.2	Symmetric Lévy Law I	116
	11.3	Uniform Random Scattering	117
	11.4	One-Sided Lévy Law II	118
	11.5	Symmetric Lévy Law II	119
	11.6	Summary	120
	11.7	Methods	120
		11.7.1 Equations (11.2) and (11.3)	120
		11.7.2 Equations (11.5) and (11.6)	121
	References	123
12	**Dynamics**	125
	12.1	Growth and Decay	125
	12.2	Evolution	126
	12.3	Vanishing and Exploding	127
	12.4	Order Statistics	128
	12.5	Beyond the Singularity	129

		12.6	Summary	130
		12.7	Methods	130
		References		131
13	**Limit Laws**			133
	13.1	Limit Laws I		133
	13.2	Limit Laws II		134
	13.3	Limit Laws III		136
	13.4	Limit Laws IV		137
	13.5	Limit Laws V		139
	13.6	Outlook		140
	13.7	Methods		143
		13.7.1	Preparation I	143
		13.7.2	The Limit $\mathcal{E}_+$ via Eq. (13.5)	145
		13.7.3	The Limit $\mathcal{E}_-$ via Eq. (13.7)	145
		13.7.4	The Limit $\mathcal{E}_+$ via Eq. (13.9)	146
		13.7.5	The Limit $\mathcal{E}_-$ via Eq. (13.9)	147
		13.7.6	Preparation II	148
		13.7.7	The Limit $\mathcal{E}_+$ via Eq. (13.19)	149
		13.7.8	The Limit $\mathcal{E}_-$ via Eq. (13.19)	150
	References			151
14	**First Digits**			153
	References			157
15	**Motions**			159
	15.1	Aggregation		159
	15.2	Diffusive Motions		160
	15.3	Regular Diffusion		161
	15.4	Anomalous Diffusion		161
	15.5	Stationary Velocities		163
	15.6	White Noise		164
	15.7	Flicker Noise		165
	15.8	Fusion		166
	15.9	Outlook		167
	15.10	Methods		169
		15.10.1	Equation (15.4)	169
		15.10.2	Invariance and Eq. (15.7)	170
		15.10.3	Anomalous-Diffusion Examples	170
		15.10.4	Equation (15.10)	172
		15.10.5	Invariance and Eq. (15.13)	174
		15.10.6	Equation (15.14)	174
	References			175

16	**First Passage Times**	177
	References	181
17	**From Power to Lognormal**	183
	17.1 Double-Pareto Laws	183
	17.2 Langevin and Gibbs	185
	17.3 Exponentiation	186
	17.4 U-Shaped Potentials	187
	17.5 Edge of Convexity	189
	17.6 Universal Approximation	191
	17.7 Lognormal and Log-Laplace Scenarios	192
	17.8 Summary	193
	17.9 Methods	194
	17.9.1 Equation (17.8)	194
	17.9.2 Equations (17.14) and (17.15)	194
	References	196
18	**Conclusion**	199

Notation

General Notation

IID	is acronym for "Independent and Identically Distributed" (random variables)
E[X]	is the expectation of a (real-valued) random variable X
Var[X]	is the variance of a (real-valued) random variable X
Cov[X, Y]	is the covariance between two (real-valued) random variables X and Y

The Power Poisson Process $\mathcal{E}_+$

$\lambda_+(x) = c\epsilon x^{\epsilon-1}\ (x>0)$	is the intensity function; the coefficient c and the exponent ϵ are positive parameters
$\Lambda_+(l) = cl^\epsilon\ (l>0)$	is the mean number of points residing below the level l
$N_+(l)$	is the number of points residing below the level l
$B_+(l)$	is the position of a "representative point" residing below the level l
$A_+(l)$	is the position of the point that is closest, from above, to the level l
$A_+(0)$	is the position of the smallest point (i.e., the minimal point)
$P_+(k)$	is the position of the kth order statistic ($k = 1, 2, 3, \ldots$)

The Power Poisson Process $\mathcal{E}_-$

$\lambda_-(x) = c\epsilon x^{-\epsilon-1}\ (x>0)$	is the intensity function; the coefficient c and the exponent ϵ are positive parameters

$\Lambda_-(l) = cl^{-\epsilon}$ $(l > 0)$	is the mean number of points residing above the level l
$N_-(l)$	is the number of points residing above the level l
$A_-(l)$	is the position of a "representative point" residing above the level l
$B_-(l)$	is the position of the point that is closest, from below, to the level l
$B_-(\infty)$	is the position of the largest point (i.e., the maximal point)
$P_-(k)$	is the position of the kth order statistic ($k = 1, 2, 3, \ldots$)

Chapter 1
Introduction

Power laws are prevalent in the physical sciences, e.g., Newton's law of gravitation; Coulomb's law of electrostatics; Kepler's third law of planetary motion; the Stefan–Boltzmann law of black-body radiation; and the inverse-square law of radiation decay. These examples demonstrate precise deterministic power relations between inputs and outputs. However, as our world is abundant with systems that are both complex and noisy, it is not always possible to establish precise deterministic relations between inputs and outputs. Consequently, the investigations of complex and noisy systems often involve statistical methods, i.e.: empirical data is collected and then represented in statistical forms. There are various ways of representing empirical data statistically, and each way gives rise to a different type of a statistical power law. In this monograph, we shall address an assortment of statistical power laws, and we shall henceforth term such laws *power statistics*.

Let's begin with two principal statistical forms: distribution functions and rank distributions. To that end consider the empirical data to be a sample of values that were collected from some positive quantity of interest. The *tail distribution function* describes the proportion of the values that are greater than a given threshold level; in this statistical form the threshold level is the input, and the corresponding proportion is the output. The *rank distribution* orders the values decreasingly: the largest value is assigned rank one, the second largest value is assigned rank two, the third largest value is assigned rank three, etc.; in this statistical form the rank is the input, and the corresponding value is the output.

A power relation between the input and output of a tail distribution function is termed *Pareto's law* [1, 2], in honor of the Italian economist Vilfredo Pareto. Analyzing tax data, Pareto observed that such a power relation holds for the income and wealth of the rich [3]; apparently, Pareto's observation was the first documented discovery of power statistics. A power relation between the input and output of a rank distribution is termed *Zipf's law* [4], in honor of the American linguist George Kingsley Zipf [5, 6]. In linguistics such a power relation holds for the frequencies of different words appearing in a large text; as shall be evident from the references hereinafter, Zipf was *not* the first to discover this type of power statistics.

© Springer Nature Switzerland AG 2020
I. Eliazar, *Power Laws*, Understanding Complex Systems,
https://doi.org/10.1007/978-3-030-33235-8_1

Another key statistical form is the *Lorenz curve*, named in honor of the American economist Max Lorenz [7]. The Lorenz curve is widely applied in economics and in the social sciences to quantify, in a canonical manner, the distribution of income and wealth in human societies [8–10]. Specifically, for a human society under consideration: the Lorenz curve describes the proportion of the society's overall wealth that is held by the society members that are above a given percentile; in this statistical form the percentile is the input, and the corresponding proportion is the output. The Lorenz curve combines together elemental features of both distribution functions and rank distributions, and it is applicable in the context of non-negative quantities at large [11–13]. A power relation between the input and output of a Lorenz curve manifests Pareto's law [14]. Also, on a macroscopic scale, Zipf's law yields a Lorenz curve with a power relation between its input and output [15, 16].

The scientific interest in Pareto's law and in Zipf's law is vast and transdisciplinary [17–39]. Specific examples of empirical data displaying these laws include: income and wealth [3, 40–44]; city sizes [45–49]; word frequencies [50–54]; scientific productivity [55–59]; firm sizes [60–64]; self-organized criticality [65–68]; and degree distributions in networks [69–73].[1] Note that in these examples the underlying real-world systems are highly complex and noisy indeed. Pareto's law and Zipf's law are the paradigmatic types of power statistics appearing in the literature. Yet, the buck does *not* stop at Pareto and Zipf.

In reliability engineering, statistical distributions are often represented in the form of *hazard rates* [74, 75]. The hazard rate of the lifespan of a system of interest has the following meaning: given that the system has operated till a certain time epoch, the hazard rate is the likelihood that the system will fail right at this time epoch. A power hazard rate characterizes the *Weibull law*, which is named in honor of the Swedish scholar Waloddi Weibull [76, 77]. The Weibull law is one of the three *extreme-value laws* that emerge universally from the Fisher–Tippett–Gnedenko theorem [78, 79], a keystone result of Extreme-Value Theory [80–84]. The Weibull law has numerous applications in science and engineering [85–90]. In particular, a special case of the Weibull law—commonly termed *stretched exponential*—is an archetypal model for *anomalous relaxation* [91–97].

In statistical physics and in probability theory, statistical distributions are often represented via integral transforms. Specifically: the statistical distribution of a non-negative quantity is represented by its *Laplace transform*; and the statistical distribution of a real quantity is represented by its *Fourier transform*. An inherent power-structure of these integral transforms characterizes, respectively, the one-sided and the symmetric Lévy laws. These laws belong to the class of *Lévy-stable laws*, which are named in honor of the French mathematician Paul Lévy [98]. The Lévy-stable laws emerge from the generalized Central Limit Theorem [99–101], and are of profound importance in science and engineering [102–109]. In particular, the one-sided and the symmetric Lévy laws play key roles in the topics of *Lévy walks* and *Lévy flights* [110–116].

---

[1] As noted above, Zipf was not the first to discover the law named after him: this law was first discovered by Auerbach [45] (city sizes), then by Estoup [50] (word frequencies), and then by Lotka [55] (scientific productivity).

1 Introduction

Let's return now to the sample of positive values from which the aforementioned distribution functions and rank distributions were constructed. These values are commonly represented via the decimal numerical system. So, let's address the *first digits* of these values; for example, the first digit of the number 735 is 7, the first digit of the number 0.005412 is 5, and the first digit of the number $1/3 = 0.333 \cdots$ is 3. Plain intuition suggests that, for a broad sample of values, the first digits will display uniform statistics, i.e.: each of the digits $\{1, 2, \cdots, 9\}$ will appear with frequency close to $1/9$. Surprisingly, the reality is quite different from plain intuition.

For a broad sample of values, the first digits often display a *logarithmic* statistical pattern. This pattern is termed *Newcomb–Benford's law* [117, 118], in honor of the Canadian-American astronomer Simon Newcomb [119] and the American engineer Frank Benford [120]. The Newcomb–Benford law attracted significant attention, popular and scientific alike [121–129]. Interestingly, the Newcomb–Benford law is an effective tool in forensic accounting, auditing, and fraud detection [130]. Recently, it was established that the Newcomb–Benford law is a special case of a more general law [131]: a *power* statistical pattern for first digits.

Up to this point, we addressed one-dimensional quantities—positive, non-negative, real, and integer. Elevating from one-dimensional quantities to infinite-dimensional objects we arrive at processes, and two principal types of such stochastic objects are: *diffusions* and *stationary processes*. On the one hand, diffusions model random motions that diffuse as time progresses. On the other hand, stationary processes model random motions whose behavior is stationary with respect to the progress of time. Both these types of random motions are omnipresent in science and engineering [132–134].

The most widely applied quantitative tool in the context of diffusions is tracking their Mean Square Displacement (MSD) over time. Specifically, consider a diffusive random motion that initiates at time zero, and set its temporal MSD function as follows: at any given (non-negative) time epoch, the function's value is the MSD of the motion's position at that time epoch; the displacement is measured relative to the motion's initial position. A diffusion with a linear temporal MSD function is considered "regular". A diffusion with a power temporal MSD function is termed *sub-diffusion* when the function is sub-linear, and is termed *super-diffusion* when the function is super-linear. Sub-diffusions and super-diffusions are main pillars of the field of *anomalous diffusion* [135–148].

The correlation structure of a stationary process is characterized by its auto-covariance function. In turn, the Fourier transform of the auto-covariance function is the *spectral density* of the stationary process under consideration [149–151]. A stationary process with a power spectral density is termed *flicker noise*; in particular, a stationary process with a harmonic spectral density is termed *1/f noise*. As in the case of Pareto's law and Zipf's law, the scientific interest in flicker noise and 1/f noise is vast and transdisciplinary [152–175].

So, as described above, there is an assortment of power statistics that are represented via various statistical forms: distribution functions, rank distributions, Lorenz curves, hazard rates, Laplace transforms, Fourier transforms, first-digit distributions, temporal MSD functions, and spectral densities. In fact, as we shall see along this monograph, there are even more power statistics. This "zoo" of power statistics gives rise to the following question: **Is there a foundational statistical power-structure that underlies all the aforementioned power statistics?** The answer, as we shall argue in detail in this monograph is affirmative: a statistical 'bedrock' power-structure does exist indeed. To introduce and explore this bedrock power-structure, we need to transcend from the realm of *random variables* [176, 177] to the realm of *Poisson processes* [178–182].

Random variables and Poisson processes are two different statistical modeling methods. Consider a collection of points that are scattered randomly over a general space. The method of random variables is an *individual* approach: it quantifies the statistics of a *representative point* of the collection—doing so by picking, at random, a single point of the collection. It is important to note that if the collection comprises of infinitely many points then picking a single point at random is *not* possible, and hence the collection has *no* "representative point". On the other hand, the method of Poisson processes is a *collective* approach: it quantifies the statistics of the *entire collection*.

Poisson processes assume a major role in probability theory and in applied probability [178–182], with applications ranging from queueing systems to insurance and finance [183, 184]. To underscore the difference between the random-variables method and the Poisson-processes method let us consider, for the sake of illustration, the underlying space to be the positive half-line $(0, \infty)$.

In the first method the random variable is the position of a "representative point", and the statistics of this random variable are commonly quantified by a *probability density*: a positive-valued function, $f(x)$ $(x > 0)$, whose integral over the positive half-line has a unit mass, $\int_0^\infty f(x)\,dx = 1$. The meaning of the probability density is as follows: the probability that the random variable be realized in the infinitesimal interval $(x, x + dx)$ is $f(x)\,dx$; and if the random variable is realized in the infinitesimal interval $(x, x + dx)$ then it is *not* realized elsewhere.

In the second method, the Poisson process is the entire collection of points, and the statistics of this Poisson process are commonly quantified by a *Poisson intensity*: a positive-valued function, $\lambda(x)$ $(x > 0)$, whose integral over the positive half-line has either a finite or an infinite mass, $\int_0^\infty \lambda(x)\,dx = m$ where $0 < m \leq \infty$. The meaning of the Poisson intensity is as follows: the probability that the infinitesimal interval $(x, x + dx)$ contains a point is $\lambda(x)\,dx$; and whatever happens in the infinitesimal interval $(x, x + dx)$ is independent of whatever happens elsewhere.

As evident from their descriptions, the approaches underpinning these two modeling methods are markedly different. On the one hand, the random-variables method is applied ubiquitously in science and engineering; however, this method is essentially incapable of modeling infinite collections. On the other hand, the Poisson-processes method is well up for the task of modeling infinite collections; however, the use of this method across the sciences is limited.

The Poisson-processes method actually extends the random-variables method. Indeed, in the case of finite collections the Poisson-processes method effectively coincides with the random-variables method. When the underlying space is the positive half-line $(0, \infty)$, then the coincidence of the two methods is manifested by the following connection between the aforementioned probability density and the afore-

1 Introduction

mentioned Poisson intensity [178][2]: $f(x) = \lambda(x)/m$ $(x > 0)$; namely, the probability density $f(x)$ is the unit-mass normalization of the Poisson intensity $\lambda(x)$.

With the method of Poisson processes described, we are now in a position to state the following affirmative answer to the above question: **The power Poisson process is the foundational statistical power-structure that underlies all the aforementioned power statistics**. Specifically, the power Poisson process is a Poisson process over the positive half-line $(0, \infty)$ with a power intensity function:

$$\lambda(x) = Cx^{p-1}$$

$(x > 0)$, where the coefficient C is a positive parameter, and where the power $p \neq 0$ is either a positive or a negative parameter. The power Poisson process comprises infinitely many points, and the mass of its intensity function is infinite $\int_0^\infty \lambda(x)\,dx = \infty$.

There is a profound difference between the case of positive powers ($p > 0$), the case of negative powers ($p < 0$), and the boundary case of a zero power ($p = 0$). The boundary case—whose intensity function is harmonic, $\lambda(x) = C/x$ ($x > 0$)—was studied in detail in [185]. This monograph addresses the cases of positive and negative powers ($p \neq 0$).

Having presented the above power-Poisson-process answer, a follow-up question arises: where does this power Poisson process come from? The answer to this follow-up question, which shall also be presented in this monograph, is: *stochastic limit laws*.

Perhaps the best-known stochastic limit law in science and engineering is the Central Limit Theorem (CLT) [186, 187]. To describe the CLT, consider a large collection of statistically identical real-valued random variables. The CLT focuses on the sum of the collection's random variables, and—under certain conditions, and with respect to a specific scaling scheme—it asserts that: as the collection's size grows infinitely large, its sum converges in law to a limiting *Normal random variable* [188]. The probability density of the limiting Normal random variable is the iconic *Gauss bell curve* [188].

Analogously to the CLT, a set of stochastic limit laws that yield the power Poisson process shall be established here. To describe these stochastic limit laws consider a large ensemble of points that are scattered randomly over the positive half-line. Under certain conditions, and with respect to specific scaling schemes, the stochastic limit laws assert that: as the ensemble's size grows infinitely large, it converges in law to a limiting ensemble that is the *power Poisson process*. The *power intensity function* $\lambda(x) = Cx^{p-1}$ of the limiting power Poisson process is the counterpart of the Gauss-bell-curve probability density that emerges from the CLT.

On the one hand, the Gauss bell curve is a quintessential manifestation of "*normal*" statistical behavior: it is a universal pattern of "*mild*" randomness [189–191]. On the other hand, the power intensity function $\lambda(x) = Cx^{p-1}$ is the foundational bedrock of an assortment of "*anomalous*" statistical behaviors: power statistics that are universal patterns of "*wild*" randomness [189–191].

---

[2]In the case of finite collections, the mass m of the Poisson intensity is finite, $0 < m < \infty$.

Figuratively, the power Poisson process is an infinite iceberg, and the aforementioned power statistics are some of its tip-of-the-iceberg facets. In this monograph, we explore the iceberg and its tip facets via diverse perspectives: structural, probabilistic, fractal, dynamical, socioeconomic, and stochastic. This exploration provides an overarching panoramic view, as well as comprehensive in-depth analyses, of power statistics.

The topics of this monograph are organized in chapters as described below. Readers are invited to either read the chapters sequentially, or to hop between chapters as they please; in the latter case the readers are advised to begin with Chap. 3.

Chapter 1: Introduction.

Chapter 2: A stochastic limit law leading, in a universal fashion, from lognormal statistics to the power Poisson process.

Chapter 3: A framework for the power Poisson process, and a terse review of the Poisson law.

Chapter 4: A threshold analysis of the power Poisson process; in particular, this analysis will yield the Pareto and Weibull laws.

Chapter 5: The connection between the power Poisson process and power hazard rates.

Chapter 6: The connection between Pareto's law and a forecasting heuristic termed Lindy's law.

Chapter 7: An order-statistics analysis of the power Poisson process; in particular, this analysis will yield the Weibull and Zipf laws.

Chapter 8: The statistical estimation of the power parameter p of the power Poisson process.

Chapter 9: A socioeconomic analysis of the power Poisson process; in particular, this analysis will yield power Lorenz curves.

Chapter 10: A fractal analysis of the power Poisson process.

Chapter 11: A summation analysis of the power Poisson process; in particular, this analysis will yield the one-sided and symmetric Lévy laws.

Chapter 12: The connection between the power Poisson process and nonlinear power dynamics.

Chapter 13: Stochastic limit laws leading to the power Poisson process.

Chapter 14: A first-digits analysis of the power Poisson process; in particular, this analysis will yield the generalized Newcomb–Benford law.

Chapter 15: The universal effect of the power Poisson process on the sums of random motions; in particular, this effect will yield sub-diffusion, super-diffusion, flicker noise, and 1/f noise.

Chapter 16: The universal effect of the power Poisson process on the first passage times of collections of random motions; in particular, this effect will yield an innovation rate termed Herdan–Heaps law.

Chapter 17: Returning back from the power Poisson process to lognormal statistics —from which we set off in Chap. 2.

Chapter 18: Conclusion.

References

1. M. Hardy, Math. Intel. **32**, 38 (2010)
2. B.C. Arnold, *Pareto Distributions* (CRC Press, Boca Raton, 2015)
3. V. Pareto, Cours d'économie politique (Droz, Geneva, 1896); V. Pareto, Manual of political economy, reprint edition (Oxford University Press, Oxford, 2014)
4. A. Saichev, D. Sornette, Y. Malevergne, *Theory of Zipf's Law and Beyond* (Springer, New York, 2009)
5. G.K. Zipf, *The Psychobiology of Language* (Houghton-Mifflin, Boston, 1935)
6. G.K. Zipf, *Human Behavior and the Principle of Least Effort* (Addison-Wesley, Cambridge, 1949)
7. M.O. Lorenz, Pub. Amer. Stat. Assoc. **9**, 209 (1905)
8. J.L. Gastwirth, Econometrica **39**, 1037 (1971)
9. G.M. Giorgi, Metron **63**, 299 (2005)
10. D. Chotikapanich (ed.), *Modeling Income Distributions and Lorenz Curves* (Springer, New York, 2008)
11. I. Eliazar, I.M. Sokolov, Physica A **391**, 1323 (2012)
12. I. Eliazar, Phys. Rep. **649**(1) (2016)
13. I. Eliazar, Ann. Phys. **389**, 306 (2018)
14. I. Eliazar, M.H. Cohen, Physica A **402**, 30 (2014)
15. I. Eliazar, M.H. Cohen, Physica A **390**, 4293 (2011)
16. I. Eliazar, M.H. Cohen, Phys. Rev. E **89**, 012111 (2014)
17. J.P. Bouchaud, Quant. Fin. **1**, 105 (2001)
18. W.J. Reed, Econ. Lett. **74**, 15 (2001)
19. W.J. Reed, B.D. Hughes, Phys. Rev. E **66**, 067103 (2002)
20. Glottometrics 5 (2002); The entire issue is on Zipf's law
21. W.J. Reed, Physica A **319**, 469 (2003)
22. W.J. Reed, M. Jorgensen, Com. Stat. Theor. Meth. **33**, 1733 (2004)
23. M. Mitzenmacher, Internet Math. **1**, 226 (2004)
24. A.B. Downey, Comput. Commun. **28**, 790 (2005)
25. A. Chatterjee, S. Yarlagadda, B.K. Chakrabarti (eds.), *Econophysics of Wealth Distributions* (Springer, Milan, 2005)
26. L. Egghe (ed.), *Power Laws in the Information Production Process: Lotkaian Informetrics* (Emerald Group Publishing, Bingley UK, 2005)
27. M.E.J. Newman, Contemp. Phys. **46**, 323 (2005)
28. D. Sornette, *Critical Phenomena in Natural Sciences* (Springer, New York, 2006)
29. G. Martinez-Mekler, R.A. Martinez, M.B. del Rio, R. Mansilla, P. Miramontes, G. Cocho, PLoS ONE **4**, e4791 (2009)
30. A. Clauset, C.R. Shalizi, M.E.J. Newman, SIAM Rev. **51**, 661 (2009)
31. X. Gabaix, Annu. Rev. Econ. **1**, 255 (2009)
32. V.M. Yakovenko, J.B. Rosser, Rev. Mod. Phys. **81**, 1703 (2009)
33. M.P.H. Stumpf, M.A. Porter, Science **335**, 665 (2012)
34. M. Cristelli, M. Batty, L. Pietronero, Sci. Rep. **2**, 812 (2012)
35. B.K. Chakrabarti, A. Chakraborti, S.R. Chakravarty, A. Chatterjee, *Econophysics of Income and Wealth Distributions* (Cambridge University Press, Cambridge, 2013)
36. C.M.A Pinto, A. Mendes Lopes, J.A. Tenreiro Machado, Commun. Nonlinear Sci. Numer. Simul. **17**, 3558 (2012)
37. P. Arvanitidis, C. Kollias, Peace Econ. Peace Sci. Public Policy **22**, 41 (2016)
38. X. Gabaix, J. Econ. Persp. **30**, 185 (2016)
39. S.I. Kumamoto, T. Kamihigashi, Front. Phys. **6**, 20 (2018)
40. B. Mandelbrot, International Econ. Rev. **1**, 79 (1960)
41. H. Aoyama et al., Fractals **8**, 293 (2000)
42. O.S. Klass, O. Biham, M. Levy, O. Malcai, S. Solomon, Econ. Lett. **90**, 290 (2006)

43. C.I. Jones, J. Econ. Persp. **29**, 29 (2015)
44. S. Aoki, M. Nirei, Am. Econ. J: Macroecon. **9**, 36 (2017)
45. F. Auerbach, Petermanns Geographische Mitteilungen **59**, 74 (1913)
46. P. Krugman, J. Jap. Intern. Econ. **10**, 399 (1996)
47. X. Gabaix, Am. Econ. Rev. **89**, 129 (1999)
48. X. Gabaix, Quart. J. Econ. **114**, 739 (1999)
49. K.T. Soo, Reg. Sci. Urban Econ. **35**, 239 (2005)
50. J.B. Estoup, *Gammes Stenographiques* (Institut Stenographique de France, Paris, 1916)
51. J. Serra, A. Corral, M. Boguna, M. Haro, J.L. Arcos, Sci. Rep. **2**, 521 (2012)
52. A.M. Petersen, J.N. Tenenbaum, S. Havlin, H.E. Stanley, M. Perc, Sci. Rep. **2**, 943 (2012)
53. S.T. Piantadosi, Psychonomic Bull. Rev. **21**, 1112–1130 (2014)
54. I. Moreno-Sanchez, F. Font-Clos, A. Corral, PloS one **11**, e0147073 (2016)
55. A.J. Lotka, J. Washington Acad. Sci. **16**, 317 (1926)
56. S. Redner, Euro. Phys. J. B: Cond. Matt. Comput. Syst. **4**, 131 (1998)
57. L. Egghe, J. Assoc. Inform. Sci. Tech. **56**, 664 (2005)
58. R.A. Fairthorne, J. Documentation **61**, 171 (2005)
59. A.M. Petersen, H.E. Stanley, S. Succi, Sci. Rep. **1**, 181 (2011)
60. M.H.R. Stanley, S.V. Buldyrev, S. Havlin, R. Mantegna, M.A. Salinger, H.E. Stanley, Econ. Lett. **49**, 453 (1995)
61. R.L. Axtell, Science **293**, 1818 (2001)
62. Y. Fujiwara, Physica A **337**, 219 (2004)
63. B. Podobnik, D. Horvatic, A.M. Petersen, B. Urosevic, H.E. Stanley, Proc. Natl. Acad. Sci. (USA) **107**, 18325 (2010)
64. M. Bee, M. Riccaboni, S. Schiavo, Physica A **481**, 265 (2017)
65. P. Bak, C. Tang, J. Geophys. Res: Solid Earth **94**, 15635 (1989)
66. P. Bak, K. Chen, Sci. Am. **264**, 46 (1991)
67. B. Jiang, Int. J. Geog. Inform. Sci. **23**, 1033 (2009)
68. D. Markovic, C. Gros, Phys. Rep. **536**, 41 (2014)
69. L.A. Adamic, R.M. Lukose, A.R. Puniyani, B.A. Huberman, Phys. Rev. E **64**, 046135 (2001)
70. L.A. Adamic, B.A. Huberman, Glottometrics **3**, 143 (2002)
71. R. Albert, A.L. Barabási, Rev. Mod. Phys. **74**, 47 (2002)
72. R. Cohen, S. Havlin, *Complex Networks: Structure, Robustness and Function* (Cambridge University Press, Cambridge, 2010)
73. L. Muchnik et al., Sci. Rep **3**, 1783 (2013)
74. E. Barlow, F. Proschan, *Mathematical Theory of Reliability* (SIAM, Philadelphia, 1996)
75. M. Finkelstein, *Failure Rate Modelling for Reliability and Risk* (Springer, New York, 2008)
76. W. Weibull, Ingeniors Vetenskaps Akademiens. Stockholm **151** (1939)
77. W. Weibull, ASME, J. Appl. Mech. **18**, 293 (1951)
78. R.A. Fisher, L.H.C. Tippett, Proc. Cambridge Phil. Soc. **24**, 180 (1928)
79. B. Gnedenko, Ann. Math. **44**, 423 (1943). (Translated and reprinted in: Breakthroughs in Statistics I, ed. by S. Kotz and N.L. Johnson, pp. 195–225, Springer, New York, 1992)
80. J. Galambos, *Asymptotic Theory of Extreme Order Statistics* 2nd edn (Krieger, Melbourne, 1987)
81. S. Kotz, S. Nadarajah, *Extreme Value Distributions* (Imperial College Press, London, 2000)
82. J. Beirlant, Y. Goegebeur, J. Segers, J. Teugels, *Statistics of Extremes: Theory and Applications* (Wiley, New York, 2004)
83. L. de Haan, A. Ferreira, *Extreme Value Theory: An Introduction* (Springer, New York, 2006)
84. R.D. Reiss, M. Thomas, *Statistical Analysis of Extreme Values: with Applications to Insurance, Finance, Hydrology and Other Fields* (Birkhauser, Basel, 2007)
85. D.N.P. Murthy, M. Xie, R. Jiang, *Weibull Models* (Wiley, New York, 2003)
86. B. Dodson, *The Weibull Analysis Handbook* (ASQ Press, Milwaukee, 2006)
87. R. Abernethy, *The New Weibull Handbook*, 5th edn. (Dr. Robert, Abernethy, 2006)
88. H. Rinne, *The Weibull Distribution: A Handbook* (CRC Press, Buca Raton, 2008)

References

89. J.I. McCool, *Using the Weibull Distribution: Reliability, Modeling and Inference* (Wiley, New York, 2012)
90. C.D. Lai, *Generalized Weibull Distributions* (Springer, New York, 2013)
91. G. Williams, D.C. Watts, Trans. Faraday Soc. **66**, 80 (1970)
92. R.V. Chamberlin, G. Mozurkewich, R. Orbach, Phys. Rev. Lett. **52**, 867 (1984)
93. J.C. Phillips, Rep. Prog. Phys. **59**, 1133 (1996)
94. J. Laherrere, D. Sornette, Eur. Phys. J. B **2**, 525 (1998)
95. W.T. Coffey, YuP Kalmykov, *Fractals, Diffusions and Relaxation in Disordered Systems* (Wiley-Interscience, New York, 2006)
96. Y. Yu et al., Phys. Rev. Lett. **115**, 165901 (2015)
97. A. Mignan, Geophys. Res. Lett. **42**, 9726 (2015)
98. P. Lévy, Théorie de l'addition des variables Alé
99. B.V. Gnedenko, A.N. Kolmogorov, *Limit Distributions for Sums of Independent Random Variables* (Addison-Wesley, London, 1954)
100. I.A. Ibragimov, YuV Linnik, *Independent and Stationary Sequences of Random Variables* (Walterss-Noordhoff, Groningen, 1971)
101. N.H. Bingham, C.M. Goldie, J.L. Teugels, *Regular Variation* (Cambridge University Press, Cambridge, 1987)
102. V.M. Zolotarev, *One-Dimensional Stable Distributions* (American Mathematical Society, Providence, 1986)
103. A. Janicki, A. Weron, *Simulation and Chaotic Behavior of Stable Stochastic Processes* (Marcel Dekker, New York, 1994)
104. G. Samrodintsky, M.S. Taqqu, *Stable Non-Gaussian Random Processes* (Chapman and Hall, New York, 1994)
105. C.L. Nikias, M. Shao, *Signal Processing with Alpha-Stable Distributions and Applications* (Wiley, New York, 1995)
106. C. Tsallis, Phys. World **10**, 42 (1997)
107. V.V. Uchaikin, V.M. Zolotarev, *Chance and Stability* (De Gruyter, Berlin, 1999)
108. J.P. Nolan, *Stable Distributions: Models for Heavy Tailed Data* (Birkhauser, New York, 2012)
109. M. Grabchak, *Tempered Stable Distributions* (Springer, New York, 2016)
110. A. Blumen, G. Zumofen, J. Klafter, Phys. Rev. A **40**, 3964 (1989)
111. J. Klafter, A. Blumen, G. Zumofen, M.F. Shlesinger, Physica A **168**, 637 (1990)
112. M.F. Shlesinger, G.M. Zaslavsky, U. Frisch, *Lévy Flights and Related Topics in Physics* (Springer, New York, 1995)
113. B.J. West, P. Grigolini, R. Metzler, T.F. Nonnenmacher, Phys. Rev. E **55**, 99 (1997)
114. I. Pavlyukevich, Y. Li, Y. Xu, A. Chechkin, J. Phys. A: Math. Theor. **48**, 495004 (2015)
115. Q. Guo, E. Cozzo, Z. Zheng, Y. Moreno, Sci. Rep. **6**, 37641 (2016)
116. M. Magdziarz, T. Zorawik, Phys. Rev. E **95**, 022126 (2017)
117. A.E. Kossovsky, *Benford's Law* (World Scientific, Singapore, 2014)
118. A. Berger, T.P. Hill, *An Introduction to Benford's Law* (Princeton University Press, Princeton, 2015)
119. S. Newcomb, Am. J. Math **4**, 39 (1881)
120. F. Benford, Proc. Am. Phil. Soc. **78**, 551 (1938)
121. R.A. Raimi, Sci. Am. **221**, 109 (1969)
122. H.R. Varian, Am. Stat. **26**, 65 (1972)
123. T.P. Hill, Am. Sci. **86**, 358 (1998)
124. R. Matthews, New Sci. 27–30 (1999)
125. T.P. Hill, Am. Math. Mon. **102**, 322 (1995)
126. E. Ley, Am. Stat. **50**, 311 (1996)
127. A.K. Formann, PLoS ONE **5**, e10541 (2010)
128. A. Berger, T.P. Hill, Math. Intell. **33**, 85 (2011)
129. S.J. Miller (ed.), *Benford's Law: Theory and Applications* (Princeton University Press, Princeton, 2015)

130. M. Nigrini, *Benford's Law: Applications for Forensic Accounting, Auditing, and Fraud Detection* (Wiley, Hoboken, 2012)
131. I. Eliazar, Physica A **392**, 3360 (2013)
132. C. Gardiner, *Handbook of Stochastic Methods* (Springer, New York, 2004)
133. N.G. Van Kampen, *Stochastic Processes in Physics and Chemistry* 3rd edn. (North-Holland, Amsterdam, 2007)
134. Z. Schuss, *Theory and Applications of Stochastic Processes* (Springer, New York, 2009)
135. H. Scher, M. Lax, Phys. Rev. B **7**, 4502 (1973)
136. M.F. Shlesinger, J. Stat. Phys. **10**, 421 (1974)
137. H. Scher, E.W. Montroll, Phys. Rev. B **12**, 2455 (1975)
138. J.P. Bouchaud, A. Georges, Phys. Rep. **195**, 12 (1990)
139. R. Metzler, J. Klafter, Phys. Rep. **339**, 1 (2000)
140. I.M. Sokolov, J. Klafter, Chaos **15**, 026103 (2005)
141. J. Klafter, I.M. Sokolov, Phys. World **18**, 29 (2005)
142. J.M. Sancho, A.M. Lacasta, K. Lindenberg, I.M. Sokolov, A.H. Romero, Phys. Rev. Lett. **92**, 250601 (2004)
143. J.M. Sancho, A.M. Lacasta, K. Lindenberg, I.M. Sokolov, A.H. Romero, Phys. Rev. Lett. **94**, 188902 (2005)
144. R. Klages, G. Radons, I.M. Sokolov (eds.), *Anomalous Transport: Foundations and Applications* (Wiley, New York, 2008)
145. M. Khoury, A.M. Lacasta, J.M. Sancho, K. Lindenberg, Phys. Rev. Lett. **106**, 090602 (2011)
146. I. Eliazar, J. Klafter, Ann. Phys. **326**, 2517 (2011)
147. Y.B. Lev, D.M. Kennes, C. Klockner, D.R. Reichman, C. Karrasch, Europhys. Lett. **119**, 37003 (2017)
148. M.A. Zaks, A. Nepomnyashchy, Proc. Natl. Acad. Sci. USA 201717225 (2018)
149. A.N. Shiryaev, *Probability* (Springer, New York, 1995)
150. G. Grimmett, D. Stirzaker, *Probability and Random Processes* (Oxford University Press, Oxford, 2001)
151. G. Lindgren, *Stationary Stochastic Processes* (CRC Press, New York, 2012)
152. W. Schottky, Ann. der Phys. **362**, 541 (1918)
153. C.A. Hartmann, Ann. der Phys. **65**, 51 (1921)
154. J.B. Johnson, Phys. Rev. **26**, 71 (1925)
155. W. Schottky, Phys. Rev. **28**, 74 (1926)
156. B.B. Mandelbrot, Int. Econ. Rev. **10**, 82 (1969)
157. R.F. Voss, J. Clarke, J. Acoust. Soc. USA **63**, 258 (1978)
158. A. van der Ziel, Adv. Electron. Electron. Phys. **49**, 225 (1979)
159. M.S. Kesner, Proc. IEEE **70**, 212 (1982)
160. E.W. Montroll, M.F. Shlesinger, Proc. Natl. Acad. Sci. USA **79**, 3380 (1982)
161. E.W. Montroll, M.F. Shlesinger, J. Stat. Phys. **32**, 209 (1983)
162. P. Bak, C. Tang, K. Wiesenfeld, Phys. Rev. Lett. **59**, 381 (1987)
163. P. Bak, C. Tang, K. Wiesenfeld, Phys. Rev. A **38**, 364 (1988)
164. M.B. Weissman, Rev. Mod. Phys. **60**, 537 (1988)
165. B.J. West, M.F. Shlesinger, Int. J. Mod. Phys. B **3**, 795 (1989)
166. W. Li, Phys. Rev. A **43**, 5240 (1991)
167. W. Li, Int. J. Bifurc. Chaos **2**, 137 (1992)
168. B.B. Mandelbrot, *Multifractals and 1/f Noise* (Springer, New York, 1999)
169. M. Lukovic, P. Grigolini, J. Chem. Phys. **129**, 184102 (2008)
170. P. Grigolini, G. Aquino, M. Bologna, M. Lukovic, B.J. West, Physica A **388**, 4192 (2009)
171. I. Eliazar, J. Klafter, Proc. Natl. Acad. Sci. (USA) **106**, 12251 (2009)
172. I. Eliazar, J. Klafter, Phys. Rev. E **82**, 021109 (2010)
173. G. Aquino, M. Bologna, P. Grigolini, B.J. West, Phys. Rev. Lett. **105**, 040601 (2010)
174. D.S. Dean, A. Iorio, E. Marinari, G. Oshanin, Phys. Rev. E **94**, 032131 (2016)
175. N. Leibovich, E. Barkai, Phys. Rev. E **96**, 032132 (2017)

References

176. W. Feller, *An Introduction to Probability Theory and Its Applications*, vol. 1, 2nd edn. (Wiley, New York, 1968)
177. W. Feller, *An Introduction to Probability Theory and Its Applications*, vol. 2, 2nd edn. (Wiley, New York, 1971)
178. J.F.C. Kingman, *Poisson Processes* (Oxford University Press, Oxford, 1993)
179. D.R. Cox, V. Isham, *Point Processes* (CRC, Boca Raton, 2000)
180. R.L. Streit, *Poisson Point Processes* (Springer, New York, 2010)
181. G. Peccati, M. Reitzner (eds.), *Stochastic Analysis for Poisson Point Processes* (Springer, New York, 2016)
182. G. Last, M. Penrose, *Lectures on the Poisson Process* (Cambridge University Press, Cambridge, 2017)
183. R.W. Wolff, *Stochastic Modeling and the Theory of Queues* (Prentice-Hall, London, 1989)
184. P. Embrechts, C. Kluppelberg, T. Mikosch, *Modelling Extremal Events for Insurance and Finance* (Springer, New York, 1997)
185. I. Eliazar, Ann. Phys. **380**, 168 (2017)
186. W.J. Adams, *The Life and Times of the Central Limit Theorem* (AMS, Providence, 2009)
187. H. Fischer, *A History of the Central Limit Theorem* (Springer, New York, 2010)
188. J.K. Patel, C.B. Read, *Handbook of the Normal Distribution* (Dekker, New York, 1996)
189. B.B. Mandelbrot, *Fractals and Scaling in Finance* (Springer, New York, 1997)
190. B.B. Mandelbrot, N.N. Taleb, *The Known, the Unknown and the Unknowable in Financial Institutions*, ed. by F. Diebold, N. Doherty, R. Herring (Princeton University Press, Princeton, 2010), pp. 47–58
191. I. Eliazar, M.H. Cohen, Chaos. Solitons & Fractals **74**, 3 (2015)

Chapter 2
From Lognormal to Power

In this chapter we establish, via a lognormal-statistics path, the emergence of the power Poisson process. We will show how the power Poisson process materializes from fairly general multiplicative stochastic dynamics in continuous time. This chapter is based on [1], which investigated the universal Poisson-process limits that emerge from general random walks.

2.1 Geometric Evolution

Perhaps the most basic and prevalent ordinary differential equation (ODE), in dimension one, is the *linear* ODE. Specifically, the linear ODE is given by $\dot{X}(t)/X(t) = r$ ($t \geq 0$), where: the variable t manifests time; $X(t)$ is the size, at time t, of some positive quantity of interest that is being tracked; and r is a real parameter. An alternative way of writing the linear ODE is the form

$$\frac{\dot{X}(t)}{X(t)} = r \tag{2.1}$$

($t \geq 0$). Equation (2.1) implies that the parameter r is the quantity's *geometric* rate of change, i.e., its rate of change *relative* to its size.

The solution of the linear ODE is given by the exponential

$$X(t) = X(0) \exp(rt) \tag{2.2}$$

($t \geq 0$). Hence, the linear ODE describes *geometric evolution*. The temporal behavior of the geometric evolution is determined by the *sign* of the geometric rate: exponential decay if and only if the rate is negative, $r < 0$; constant behavior if and only if the rate is zero, $r = 0$; exponential growth if and only if the rate is positive, $r > 0$.

© Springer Nature Switzerland AG 2020
I. Eliazar, *Power Laws*, Understanding Complex Systems,
https://doi.org/10.1007/978-3-030-33235-8_2

2.2 Gaussian White Noise

In the real world, *noise* is omnipresent and always present. In the context of the linear ODE (2.1) the presence of noise can be incorporated via the geometric rate. Namely, rather than considering the geometric rate to be deterministic and constant, we consider it to be stochastic and to vary with time. Setting off from the linear ODE (2.1), and adding noise to the parameter r, we arrive at the following *stochastic differential equation* (SDE):

$$\frac{\dot{X}(t)}{X(t)} = r + \sqrt{\nu} \cdot \eta(t) \qquad (2.3)$$

($t \geq 0$), where ν is a positive parameter, and where $\eta(t)$ ($t \geq 0$) is a stochastic noise process.

In the SDE (2.3), $\sqrt{\nu}$ is the magnitude of the noise, and $\eta(t)$ is the noise level at time t. The geometric rate fluctuates randomly about the parameter r, and the term $\sqrt{\nu} \cdot \eta(t)$ manifests the random fluctuation at time t. We set the stochastic process $\eta(t)$ ($t \geq 0$) to be *Gaussian white noise*, the most common type of noise in science and engineering [2–4]. Consequently, the solution to the SDE (2.3) is given by

$$X(t) = X(0) \exp\left[\mu t + \sqrt{\nu} \cdot \int_0^t \eta(t')\, dt'\right] \qquad (2.4)$$

($t \geq 0$), where $\mu = r - \frac{1}{2}\nu$.

Contrary to solving the linear ODE (2.1), solving the SDE (2.3) is *not* a straightforward task. Indeed, solving the SDE (2.3) requires the use of *Ito's formula* [5–7]. This formula is a keystone of *Ito's calculus* [5–7], a stochastic calculus for SDEs driven by Gaussian white noise [8–10]. The SDE (2.3) is the paradigmatic model, in economics and finance, for the evolution of stock prices [11–13]. In the context of stock-price modeling, the parameter r is termed "*drift*" and the parameter $\sqrt{\nu}$ is termed "*volatility*" [11–13].

2.3 Brownian Motions

The solution of the SDE (2.3), given by Eq. (2.4), incorporates three foundational stochastic processes. At the very core of Eq. (2.4) stands *Brownian motion* (BM), the running integral of Gaussian white noise: $B(t) = \int_0^t \eta(t')\, dt'$ ($t \geq 0$) [6, 7, 14]. The following affine transformation of BM yields *linear Brownian motion* (LBM): $B_{\text{LBM}}(t) = \mu t + \sqrt{\nu} \cdot B(t)$ ($t \geq 0$). And the exponentiation of LBM yields *geometric Brownian motion* (GBM): $B_{\text{GBM}}(t) = \exp[B_{\text{LBM}}(t)]$ ($t \geq 0$).

BM and LBM are the archetypal models of *diffusion* in the physical sciences [2–4]; BM models "pure" diffusion, and LBM models diffusion with "drift". The

2.3 Brownian Motions

position of LBM at time t, $B_{\text{LBM}}(t)$, is a *normal (Gauss)* random variable with mean $\mathbf{E}[B_{\text{LBM}}(t)] = \mu t$ and with variance $\mathbf{Var}[B_{\text{LBM}}(t)] = \nu t$ [14]. Consequently, for any positive time t, the random variable $X(t)$ of Eq. (2.4) admits the following representation in law:

$$X(t) = X(0) \exp\left[\mu t + \sqrt{\nu t} \cdot Z\right], \qquad (2.5)$$

where Z is a *standard normal (Gauss)* random variable, i.e. with zero mean and unit variance [15].

The moment generating function of the standard normal random variable Z is given by [15]: $\mathbf{E}\left[\exp(\theta Z)\right] = \exp(\frac{1}{2}\theta^2)$ ($-\infty < \theta < \infty$). Also, recall that $\mu = r - \frac{1}{2}\nu$. Hence, assuming that the initial position $X(0)$ and the BM $B(t)$ ($t \geq 0$) are mutually independent, computing the expectation of both sides of Eq. (2.5) yields

$$\mathbf{E}[X(t)] = \mathbf{E}[X(0)] \exp(rt) \qquad (2.6)$$

($t \geq 0$). Equation (2.6) implies that, *on average*, the solution of the SDE (2.3) coincides with the solution of the linear ODE (2.1).

Appearances, however, may be misleading. And indeed, the *average* behavior of the solution of the SDE (2.3) is markedly different from its *actual* behavior—as we shall now argue. To that end we set a threshold level $l > 1$, and consider the *first passage time* (FPT) of GBM to this level [16]: $T_l = \inf\{t \geq 0 \mid B_{\text{GBM}}(t) = l\}$. The FPT T_l is an *inverse-Gaussian* random variable [17], and its behavior is determined by the *sign* of the parameter $\mu = r - \frac{1}{2}\nu$ as follows [16, 17]: if $\mu < 0$ then $\Pr(T_l = \infty) > 0$; if $\mu = 0$ then $\Pr(T_l = \infty) = 0$ and $\mathbf{E}[T_l] = \infty$; and if $\mu > 0$ then $\mathbf{E}[T_l] < \infty$.

To illustrate the marked difference between the average behavior and the actual behavior, consider the stochastic process $X(t)$ ($t \geq 0$) to be the price process of a given stock with drift parameter r that is in the range $0 < r < \frac{1}{2}\nu$. On average, the stock's price follows exponential growth. However, *in reality*, there is a positive probability that the stock's price will *never ever* double: $\Pr(T_2 = \infty) > 0$. So, while the stock's average behavior is determined by its drift parameter r alone, the stock's actual behavior is determined by both its drift and volatility parameters via the "compound" parameter $\mu = r - \frac{1}{2}\nu$.

2.4 Gaussian Motions

As noted in the previous section, the LBM $B_{\text{LBM}}(t)$ ($t \geq 0$) features a *linear* temporal structure: its mean $\mathbf{E}[B_{\text{LBM}}(t)] = \mu t$ and its variance $\mathbf{Var}[B_{\text{LBM}}(t)] = \nu t$ are both *linear* functions of the time variable t. We now generalize matters as follows: on the one hand, we maintain the Gaussian structure of LBM; on the other hand, we do not restrict the temporal structure to be linear. This generalization is attained by replacing the LBM $B_{\text{LBM}}(t)$ ($t \geq 0$) with a general *Gaussian motion* (GM): $Z(t)$ ($t \geq 0$) [18, 19].

The position of the GM at time t, $Z(t)$, is a *normal (Gauss)* random variable with mean $\mathbf{E}[Z(t)] = m(t)$ and variance $\mathbf{Var}[Z(t)] = v(t)$. Namely, the mean $m(t)$ and the variance $v(t)$ are general temporal functions. Consequently, for any positive time t, Eq. (2.5) is replaced by the following representation in law:

$$X(t) = X(0) \exp\left[m(t) + \sqrt{v(t)} \cdot Z\right], \qquad (2.7)$$

where, as in Eq. (2.5), Z is a *standard normal (Gauss)* random variable.

The mean function $m(t)$ ($t \geq 0$) manifests the deterministic trend of the GM, and the variance function $v(t)$ ($t \geq 0$) manifests the magnitude of the random fluctuations of the GM about its trend. In what follows, we consider the mean and variance functions to initiate at the origin, $m(0) = 0 = v(0)$, and we consider the variance function to be positive-valued for all positive times: $v(t) > 0$ ($t > 0$).

Set $f_t(x)$ ($x > 0$) to denote the probability density function of the random variable $X(t)$ ($t \geq 0$). The probability density function of the standard normal random variable Z is the *Gauss bell curve*: $\frac{1}{\sqrt{2\pi}} \exp(-\frac{1}{2}z^2)$ ($-\infty < z < \infty$). In turn, assuming that the initial position $X(0)$ and the GM $Z(t)$ ($t \geq 0$) are mutually independent, Eq. (2.7) implies that

$$f_t(x) = \frac{1}{\sqrt{2\pi v(t)}} \frac{1}{x} \int_0^\infty f_0(y) \exp\left\{-\frac{1}{2}\left[\frac{\ln(x/y) - m(t)}{\sqrt{v(t)}}\right]^2\right\} dy \qquad (2.8)$$

($x > 0$); the derivation of Eq. (2.8) is detailed in the Methods.

2.5 Gaussian Mean and Variance

A key feature of diffusions is that they are *diffusive*. Namely, the random fluctuations of a given diffusion motion grow larger and larger with time, and they diverge in the large-time limit $t \to \infty$. Henceforth, we assume that the general GM $Z(t)$ ($t \geq 0$) is diffusive. Specifically, we assume that the GM's variance function is asymptotically divergent:

$$\lim_{t \to \infty} v(t) = \infty. \qquad (2.9)$$

Also, we henceforth assume that the GM's mean and variance functions are asymptotically equivalent:

$$\lim_{t \to \infty} \frac{m(t)}{v(t)} = p, \qquad (2.10)$$

where p is a real limit. In other words, we assume that the GM's mean-to-variance ratio $m(t)/v(t)$ is asymptotically constant.

2.5 Gaussian Mean and Variance

In the next section, we shall discuss GMs that satisfy the assumptions of Eqs. (2.9) and (2.10). In particular, the solution of the SDE (2.3) satisfies these assumptions. Indeed, set the GM to be LBM: $Z(t) = B_{\text{LBM}}(t)$ ($t \geq 0$). Then, both the mean function and the variance function are linear: $m(t) = \mu t$ and $v(t) = \nu t$ ($t \geq 0$). Consequently, the variance function is asymptotically divergent. Also, the mean-to-variance ratio $m(t)/v(t)$ is constant for all positive times, and this constant ratio is given by

$$p = \frac{\mu}{\nu} = \frac{r}{\nu} - \frac{1}{2}. \quad (2.11)$$

Namely, in the case of the SDE (2.3), the limit parameter p is determined by the SDE's drift and the volatility parameters, r and ν.

2.6 Gaussian Stationary Velocities

In the previous section, we examined a particular GM: the LBM $B_{\text{LBM}}(t)$ ($t \geq 0$), which satisfies Eq. (2.10). In this section, we consider a GM with a *stationary velocity*—i.e., $\dot{Z}(t)$ ($t \geq 0$) is a stationary stochastic process—and determine precisely when the GM satisfies Eq. (2.10).

The covariance structure of the GM's velocity is given by

$$\mathbf{Cov}\left[\dot{Z}(t_1), \dot{Z}(t_2)\right] = \rho(|t_1 - t_2|) \quad (2.12)$$

($t_1, t_2 \geq 0$), where $\rho(|\Delta|)$ ($-\infty < \Delta < \infty$) is the velocity's auto-covariance function. Equation (2.12) means that the covariance of the velocities at times t_1 and t_2—i.e., the covariance of the random variables $\dot{Z}(t_1)$ and $\dot{Z}(t_2)$—depends only on the time-lag $|t_1 - t_2|$ between the two times.

Equation (2.12) implies that the limiting mean-to-variance ratio of Eq. (2.10) is given by

$$p = \lim_{t \to \infty} \frac{\dot{m}(t)}{2 \int_0^t \rho(t')\, dt'} = \lim_{t \to \infty} \frac{\ddot{m}(t)}{2\rho(t)} \quad (2.13)$$

provided that one of the limits appearing in Eq. (2.13) exists; the derivation of Eq. (2.13) is detailed in the Methods. The application of Eq. (2.13) is best illustrated for non-negative correlations, $\rho(t) \geq 0$ ($t \geq 0$). For such correlations the auto-covariance function can be either integrable or non-integrable: $\int_0^\infty \rho(t)\, dt < \infty$ or $\int_0^\infty \rho(t)\, dt = \infty$.

The integrable scenario ($\int_0^\infty \rho(t)\, dt < \infty$) manifests a *short-range dependence* of the stationary velocity [20–23], and in this scenario the middle part of Eq. (2.13) implies that: the velocity of the mean function, $\dot{m}(t)$, should be asymptotically constant (as $t \to \infty$); in other words, the mean function $m(t)$ should be asymptotically linear (as $t \to \infty$). The quintessential example of this scenario is *Ornstein–*

Uhlenbeck velocity $\dot{Z}(t)$ ($t \geq 0$), which is characterized by an *exponential* auto-covariance function $\rho(t)$ [24, 25].

The non-integrable scenario ($\int_0^\infty \rho(t)\,dt = \infty$) manifests *long-range dependence* of the stationary velocity [20–23], and in this scenario the right-hand part of Eq. (2.13) implies that: the acceleration of the mean function, $\ddot{m}(t)$, should be asymptotically equivalent (as $t \to \infty$) to the auto-covariance function $\rho(t)$. The quintessential example of this scenario is *fractional Brownian motion* $Z(t)$ ($t \geq 0$), whose stationary velocity is characterized by a *power* auto-covariance function $\rho(t)$ [26, 27]; in turn, the diffusion structure of fractional Brownian motion is characterized by a *power* variance function $v(t)$.

2.7 Poisson-Process Limit

Consider now an ensemble of n independent and identically distributed (IID) copies of the random variable $X(t)$ of Eq. (2.7):

$$\mathcal{E}_n(t) = \{X_1(t), \cdots, X_n(t)\}. \tag{2.14}$$

We focus our interest on the statistical behavior of the ensemble $\mathcal{E}_n(t)$ in the joint large-number and large-time limits: $n \to \infty$ and $t \to \infty$. Evidently, the first question that arises is: does such a limiting statistical behavior exist? And if the answer to the first question is positive then the second question that arises is: what are the characteristics of the limiting statistical behavior? In this section we address both these questions.

It turns out that the "goldilocks condition" that ensures a positive answer to the first question is the existence of the following limit:

$$\lim_{n,t \to \infty} n \cdot f_t(x) = \lambda(x) \tag{2.15}$$

($x > 0$), where $f_t(x)$ is the probability density function of the random variable $X(t)$ (given by Eq. (2.8)), and where $\lambda(x)$ is a positive-valued limit function. Indeed, if the "goldilocks condition" of Eq. (2.15) is met then the answer to the first question is affirmative; moreover, the limit function $\lambda(x)$ is the answer to the second question. Specifically, meeting the "goldilocks condition" of Eq. (2.15) yields the following *Poisson-process limit-law* result [1]:

▶ The ensemble $\mathcal{E}_n(t)$ converges in law, as $n \to \infty$ and $t \to \infty$, to a limiting ensemble $\mathcal{E}$ that is a *Poisson process* over the positive half-line with *intensity function* $\lambda(x)$ ($x > 0$).

The proof of this result is given in the Methods. A *Poisson process* is a countable collection of points that are scattered randomly over its domain, and whose statistics are determined by its *intensity function*—a positive-valued function that is defined

2.7 Poisson-Process Limit

over its domain [28–32]. The Poisson-process statistics are characterized by the two following properties. (I) The number of points residing in a given sub-domain is a *Poisson* random variable with mean that is given by the integral of the intensity function over the sub-domain. (II) The numbers of points residing in disjoint sub-domains are independent random variables.

A careful analysis of the probability density function $f_t(x)$ of Eq. (2.8) implies that the "goldilocks condition" of Eq. (2.15) is met if and only if the number n and the time t are coupled by the following asymptotic connection:

$$\lim_{n,t \to \infty} \frac{n}{\sqrt{2\pi v(t) \exp\left[\frac{m(t)^2}{v(t)}\right]}} = \omega, \quad (2.16)$$

where ω is a positive limit; the derivation of Eq. (2.16) is detailed in the Methods. The assumptions of Eqs. (2.9) and (2.10) ensure that the denominator appearing on the left-hand side of Eq. (2.16) is asymptotically divergent in the large-time limit $t \to \infty$. Consequently, Eq. (2.16) indeed couples the number n and the time t in the joint large-number and large-time limits, $n \to \infty$ and $t \to \infty$.

Further analysis of the probability density function $f_t(x)$ of Eq. (2.8) implies that if the asymptotic connection of Eq. (2.16) holds then the corresponding limit function of Eq. (2.15) is given by

$$\lambda(x) = \omega \mathbf{E}[X(0)^{-p}] \cdot x^{p-1} \quad (2.17)$$

($x > 0$), where ω is the positive limit of Eq. (2.16), and where p is the real limit of Eq. (2.10); the derivation of Eq. (2.17) is detailed in the Methods. The term $\mathbf{E}[X(0)^{-p}]$ appearing on the right-hand side of Eq. (2.17) is the moment of order $-p$ of the random variable $X(0)$.

2.8 Outlook

In this chapter, we constructed an *input*-to-*output* stochastic map that is described as follows. First, we set off from an initial positive-valued random variable $X(0)$. Second, we follow a geometric stochastic evolution in time:

$$X(0) \mapsto X(t) = X(0) \exp[Z(t)], \quad (2.18)$$

where $Z(t)$ ($t \geq 0$) is a Gaussian motion that is independent of the initial random variable $X(0)$. Third, we track simultaneously n IID copies of the geometric stochastic evolution of Eq. (2.18) via the ensemble

$$\mathcal{E}_n(t) = \{X_1(t), \cdots, X_n(t)\}. \quad (2.19)$$

Fourth, we focus on the limiting ensemble, in law, that emerges in the joint large-number and large-time limits: $\mathcal{E} = \lim_{n,t\to\infty} \mathcal{E}_n(t)$. The initial ensemble $\mathcal{E}_n(0)$ is the *input*, the limiting ensemble $\mathcal{E}$ is the *output*, and the Gaussian motion is the *propagator* of the input-to-output map.

The admissible class of inputs $\mathcal{E}_n(0)$ comprises of IID copies of a random variable $X(0)$ with convergent moments: $\mathbf{E}[X(0)^q] < \infty$ ($-\infty < q < \infty$). The admissible class of propagators comprises of Gaussian motions, $Z(t)$ ($t \geq 0$), whose mean and variance functions satisfy Eqs. (2.9) and (2.10). The "goldilocks coupling" of the number n and the time t is given by the asymptotic connection of Eq. (2.16). With admissible inputs and propagators, and with the "goldilocks coupling" of Eq. (2.16), the output $\mathcal{E}$ is the *power Poisson process*. Namely, $\mathcal{E}$ is a Poisson process over the positive half-line with a *power intensity function*:

$$\lambda(x) = Cx^{p-1} \tag{2.20}$$

($x > 0$), where the coefficient C is a positive parameter, and where the power p is a real parameter.

The power p is determined by the propagator, via its limiting mean-to-variance ratio of Eq. (2.10). Equation (2.17) implies that $C = \omega \mathbf{E}[X(0)^{-p}]$, and hence this coefficient is determined by three factors: (i) the random variable $X(0)$, via its moments; (ii) the power p; and (iii) the "goldilocks coupling", via the limit ω of Eq. (2.16). In effect, given the random variable $X(0)$ and the power p, we can always set the value of the coefficient C by tuning the limit ω. Hence, the intensity function of the output $\mathcal{E}$ is effectively determined by only two factors: the propagator and the "goldilocks coupling". So, the power intensity function of Eq. (2.20) is essentially *invariant* with respect to the random variable $X(0)$. Namely, the output ensemble $\mathcal{E}$ emerges in a *universal* fashion—as it does not depend on the input ensemble $\mathcal{E}_n(0)$.

As the propagator $Z(t)$ ($t \geq 0$) is a Gaussian motion, the stochastic process $\exp[Z(t)]$ ($t \geq 0$) is a geometric Gaussian motion. For any time t, the real-valued random variable $Z(t)$ is *normal (Gauss)*. Consequently, for any time t, the positive-valued random variable $\exp[Z(t)]$ is *lognormal*. The goal of this chapter is thus attained: establishing—via a lognormal-statistics path—the emergence of the power Poisson process $\mathcal{E}$. The lognormal-statistics structure underlying the geometric stochastic evolution of Eq. (2.18) played a focal role in attaining the goal, and hence the title of this chapter: "*From lognormal to power*".

2.9 Notes

We conclude with notes regarding the stochastic processes encountered along this chapter, as well as notes regarding lognormal statistics. These notes give a glimpse of the profound scientific and intellectual efforts that were invested in studying and exploring these topics.

2.9 Notes

Brownian motion (BM) is named in honor of Sir Robert Brown, who detected (1827) this motion when experimenting with pollen particles suspended in liquid [33]. However, BM was apparently first discovered (1784) by Jan Ingen-Housz when experimenting with charcoal particles suspended in liquid [34, 35]. Both Ingen-Housz and Brown observed highly erratic trajectories of the suspended particles that they tracked. Albert Einstein and Marian Smoluchowski were the first to apply (1905–1906) BM as a model of diffusion [36, 37]. Jean Perrin verified (1908–1909) experimentally the Einstein–Smoluchowski theory [38–40]. Following this theory and its experimental verification, BM assumed its predominant role as the paradigmatic model of diffusion in the physical sciences [2–4].

Norbert Wiener constructed (1923) a rigorous mathematical model of BM [41]; in mathematics and statistics, BM is often termed "Wiener process". Kiyosi Ito devised (1940s) a stochastic calculus for BM-based stochastic processes [42]. Ito's calculus is of critical importance, as it facilitates the analysis of stochastic differential equations driven by Gaussian white noise [8–10]. Donsker established (1951) that, on the macroscale, BM is the universal limit of a broad class of micro-scale random walks [43]; hence the paramount significance of this motion. BM is a random fractal object that displays a host of remarkable mathematical and statistical properties [6, 7, 14], e.g., the trajectories of BM are continuous [44], and yet they are nowhere differentiable [45].

Paul Langevin devised (1908) a stochastic framework for modeling diffusion [46]; this framework is termed "Langevin equation" [24]. Based on the Langevin equation, George Uhlenbeck and Leonard Ornstein presented (1930) a stationary model for the velocity of diffusion [47]; this model is termed "Ornstein-Uhlenbeck (OU) process". The velocity of BM is Gaussian white noise, and the auto-covariance of this noise is a delta function. On the other hand, the auto-covariance of the OU process is an exponential function. Pushing the envelope further, the fractional Brownian motion (FBM) model was introduced [48–51]; in this model, the velocity is a stationary stochastic process whose auto-covariance is a power function. The velocities of the aforementioned models exhibit the following temporal correlations: BM—zero-range; OU—short-range; FBM—long-range.

Louis Bachelier was the first to apply (1900) BM as a model for the random up-and-down movement of stock prices [52]; notably, Bachelier applied BM five years before Einstein and Smoluchowski. As BM is a real-valued stochastic process, whereas stock prices are positive-valued, Bachelier's BM model required modification. It took some sixty-five years till Paul Samuelson replaced (1965) BM with GBM as a proper model of stock prices [53]. Samuelson's GBM model stands at the core of financial option-pricing—a widely used financial framework which was devised (1973) by Robert Merton, Fisher Black, and Myron Scholes [54, 55].

Normal statistics are the principal feature of Gaussian motions—which, in particular, include the aforementioned diffusion models: BM, OU velocity, and FBM. In turn, lognormal statistics are the principal feature of geometric Gaussian motions—which, in particular, include the "exponentiations" of the aforementioned diffusion models. Lognormal statistics were pioneered (1879) by Sir Francis Galton [56]. Robert Gibrat introduced (1931) the "law of proportionate effects", which postulates that the growth rate of a firm is independent of the firm's size [57]. Combined together with the Central Limit Theorem of probability theory [58, 59], Gibrat's law universally yields lognormal statistics [60–62]. In general, lognormal statistics emerge naturally in the context of multiplicative processes [63–66]—and hence the key role that lognormal statistics play across the sciences [67–71].

As noted above, both lognormal statistics and power statistics are encountered prevalently. In the literature these two types of statistics are usually addressed separately (even when appearing in the same paper, e.g., [67, 68]). However, as established in [1], and as described in this chapter, there is a profound connection between these two seemingly separate statistics: a path leads, via a Poisson-process limit-law, from lognormal statistics to power Poisson processes—which are the "bedrock" of power statistics.

2.10 Methods

2.10.1 Equation (2.8)

Setting off from Eq. (2.7), fix a positive time t and consider the positive-valued random variable

$$U = \exp\left[m(t) + \sqrt{v(t)} \cdot Z\right], \tag{2.21}$$

where $m(t)$ is a real mean, $v(t)$ is a positive variance, and Z is a standard normal (Gauss) random variable. Denote by

$$\varphi(z) = \frac{1}{\sqrt{2\pi}} \exp\left(-\frac{1}{2}z^2\right) \tag{2.22}$$

($-\infty < z < \infty$) the probability density function of the standard normal (Gauss) random variable Z. Also, recall that $f_t(x)$ ($x > 0$) denotes the probability density function of the random variable $X(t)$, and that $f_0(x)$ ($x > 0$) denotes the probability density function of the random variable $X(0)$.

With the random variable U of Eq. (2.21), Eq. (2.7) can be written in the form $X(t) = X(0) \cdot U$. Also, the assumption that the initial position $X(0)$ and the Gaussian motion $Z(t)$ ($t \geq 0$) are mutually independent (with regard to Eq. (2.7)) translates to: the random variables $X(0)$ and U are mutually independent. Hence, we obtain that

$$\Pr[X(t) \leq x] = \Pr\left[U \leq \tfrac{x}{X(0)}\right]$$

$$= \int_0^\infty \Pr\left[U \leq \tfrac{x}{y} | X(0) = y\right] f_0(y)\,dy \tag{2.23}$$

$$= \int_0^\infty \Pr\left(U \leq \tfrac{x}{y}\right) f_0(y)\,dy$$

2.10 Methods

($x > 0$). Equation (2.21) implies that

$$\Pr\left(U \leq \tfrac{x}{y}\right) = \Pr\left[\ln(U) \leq \ln\left(\tfrac{x}{y}\right)\right]$$
$$= \Pr\left[m(t) + \sqrt{v(t)} \cdot Z \leq \ln\left(\tfrac{x}{y}\right)\right] \quad (2.24)$$
$$= \Pr\left[Z \leq \tfrac{\ln(x/y) - m(t)}{\sqrt{v(t)}}\right]$$

($x, y > 0$). Substituting Eq. (2.24) into Eq. (2.23) yields

$$\Pr[X(t) \leq x] = \int_0^\infty \Pr\left[Z \leq \frac{\ln(x/y) - m(t)}{\sqrt{v(t)}}\right] f_0(y)\, dy \quad (2.25)$$

($x > 0$). Differentiating both sides of Eq. (2.25) with respect to the variable x further yields

$$f_t(x) = \int_0^\infty \varphi\left[\frac{\ln(x/y) - m(t)}{\sqrt{v(t)}}\right] \frac{1}{x\sqrt{v(t)}} f_0(y)\, dy \quad (2.26)$$

($x > 0$). Combining together Eqs. (2.22) and (2.26) we obtain Eq. (2.8):

$$f_t(x) = \frac{1}{\sqrt{2\pi v(t)}} \frac{1}{x} \int_0^\infty f_0(y) \exp\left\{-\frac{1}{2}\left[\frac{\ln(x/y) - m(t)}{\sqrt{v(t)}}\right]^2\right\} dy \quad (2.27)$$

($x > 0$).

2.10.2 Equation (2.13)

The Gaussian motion $Z(t)$ ($t \geq 0$) initiates at the origin and hence

$$Z(t) = \int_0^t \dot{Z}(t')\, dt' \quad (2.28)$$

($t > 0$). Equation (2.28) implies that

$$\mathbf{Var}[Z(t)] = \mathbf{Cov}[Z(t), Z(t)]$$
$$= \mathbf{Cov}\left[\int_0^t \dot{Z}(t_1)\, dt_1, \int_0^t \dot{Z}(t_2)\, dt_2\right] \quad (2.29)$$
$$= \int_0^t \int_0^t \mathbf{Cov}\left[\dot{Z}(t_1), \dot{Z}(t_2)\right] dt_1 dt_2$$

($t > 0$). Using the notation $\mathbf{Var}[Z(t)] = v(t)$ and substituting Eq. (2.12) into Eq. (2.29), we obtain that

$$v(t) = \int_0^t \int_0^t \rho(|t_1 - t_2|)\, dt_1 dt_2 \tag{2.30}$$

($t > 0$). Differentiating Eq. (2.30) with respect to the time variable t yields

$$\dot{v}(t) = 2 \int_0^t \rho(t')\, dt' \tag{2.31}$$

($t > 0$). Differentiating Eq. (2.31) with respect to the time variable t further yields

$$\ddot{v}(t) = 2\rho(t) \tag{2.32}$$

($t > 0$).

Now, consider the assumptions of Eqs. (2.9) and (2.10). These assumptions call for the application of L'Hospital's rule. Doing so, and using Eq. (2.31), we obtain

$$\lim_{t \to \infty} \frac{m(t)}{v(t)} = \lim_{t \to \infty} \frac{\dot{m}(t)}{\dot{v}(t)} = \lim_{t \to \infty} \frac{\dot{m}(t)}{2 \int_0^t \rho(t')\, dt'}, \tag{2.33}$$

provided that the limit appearing on the right-hand side of Eq. (2.33) exists. If $\lim_{t \to \infty} \dot{v}(t) = \infty$ then we apply L'Hospital's rule yet again. Doing so, and using Eq. (2.32), we obtain

$$\lim_{t \to \infty} \frac{\dot{m}(t)}{\dot{v}(t)} = \lim_{t \to \infty} \frac{\ddot{m}(t)}{\ddot{v}(t)} = \lim_{t \to \infty} \frac{\ddot{m}(t)}{2\rho(t)}, \tag{2.34}$$

provided that the limit appearing on the right-hand side of Eq. (2.34) exists. Equations (2.33) and (2.34) yield Eq. (2.13).

2.10.3 A General Poisson-Process Limit-Law

Consider an ensemble $\mathcal{E}_n = \{\xi_1, \cdots, \xi_n\}$ which is a collection of n IID copies of a random variable ξ that takes values in a Euclidean domain $\mathcal{D}$. The distribution of the generic random variable ξ is characterized by a probability density function $f_t(x)$ ($x \in \mathcal{D}$) that depends on a positive parameter t. Namely: $\Pr(\xi \in D) = \int_D f_t(x)\, dx$, where D is any sub-domain of the domain $\mathcal{D}$. We assume that the probability density function $f_t(x)$ admits the form

$$f_t(x) = \frac{1}{c_t} g_t(x) \tag{2.35}$$

2.10 Methods

($x \in \mathcal{D}$) where (i) c_t is a constant satisfying

$$\lim_{t \to \infty} c_t = \infty; \tag{2.36}$$

and (ii) $g_t(x)$ ($x \in \mathcal{D}$) is a function satisfying

$$\lim_{t \to \infty} g_t(x) = g(x) \tag{2.37}$$

($x \in \mathcal{D}$), where $g(x)$ is a positive-valued function defined over the domain $\mathcal{D}$.[1]

Now, considering the large-ensemble limit $n \to \infty$ and the large-parameter limit $t \to \infty$, couple the number n and the parameter t by the following asymptotic equivalence:

$$\lim_{n,t \to \infty} \frac{n}{c_t} = \omega, \tag{2.38}$$

where ω is a positive limit value. Then, the following result holds:

Proposition 1 *The ensemble $\mathcal{E}_n$ converges in law—in the limits $n \to \infty$ and $t \to \infty$ that are coupled by Eq. (2.38)—to a limiting ensemble $\mathcal{E}$ that is a Poisson process, over the domain $\mathcal{D}$, with intensity function $\lambda(x) = \omega g(x)$ ($x \in \mathcal{D}$).*

Proof. Set $\phi(x)$ ($x \in \mathcal{D}$) to be a general, real-valued, test function that is defined over the domain $\mathcal{D}$. The characteristic functional of the ensemble $\mathcal{E}_n$ with respect to the test function $\phi(x)$ is given by

$$\mathbf{E}\left[\prod_{i=1}^{n} \phi(\xi_i)\right] = \prod_{i=1}^{n} \mathbf{E}[\phi(\xi_i)] = \mathbf{E}[\phi(\xi)]^n \tag{2.39}$$

(in Eq. (2.39) we exploited the IID structure of the ensemble $\mathcal{E}_n$).

Using the fact that $f_t(x)$ is the probability density function of the random variable ξ, and using Eq. (2.35), we have

$$\begin{aligned}
\mathbf{E}[\phi(\xi)] &= \int_{\mathcal{D}} \phi(x) f_t(x) \, dx \\
&= 1 - \int_{\mathcal{D}} [1 - \phi(x)] f_t(x) \, dx \\
&= 1 - \int_{\mathcal{D}} [1 - \phi(x)] \frac{1}{c_t} g_t(x) \, dx \\
&= 1 - \frac{1}{n} \int_{\mathcal{D}} [1 - \phi(x)] \left[\frac{n}{c_t} g_t(x)\right] dx.
\end{aligned} \tag{2.40}$$

Substituting Eq. (2.40) into Eq. (2.39) yields

$$\mathbf{E}\left[\prod_{i=1}^{n} \phi(\xi_i)\right] = \left\{1 - \frac{1}{n} \int_{\mathcal{D}} [1 - \phi(x)] \left[\frac{n}{c_t} g_t(x)\right] dx\right\}^n. \tag{2.41}$$

---

[1] We don't go into the technical mathematical details of the convergence that Eq. (2.37) describes.

Due to Eqs. (2.37) and (2.38), we have

$$\lim_{n,t\to\infty}\left[\frac{n}{c_t}g_t(x)\right]=\omega g(x) \qquad (2.42)$$

($x \in \mathcal{D}$). Setting $\lambda(x) = \omega g(x)$ ($x \in \mathcal{D}$), Eq. (2.42) implies that

$$\lim_{n,t\to\infty}\int_{\mathcal{D}}[1-\phi(x)]\left[\frac{n}{c_t}g_t(x)\right]dx = \int_{\mathcal{D}}[1-\phi(x)]\lambda(x)\,dx\,. \qquad (2.43)$$

Combined together, Eqs. (2.41) and (2.43) imply that

$$\lim_{n,t\to\infty}\mathbf{E}\left[\prod_{i=1}^{n}\phi(\xi_i)\right] = \exp\left\{-\int_{\mathcal{D}}[1-\phi(x)]\lambda(x)\,dx\right\}\,. \qquad (2.44)$$

The right-hand side of Eq. (2.44) is the characteristic functional of a Poisson process, over the domain $\mathcal{D}$, with intensity function $\lambda(x)$ ($x \in \mathcal{D}$) [28]. This proves Proposition 1. ∎

2.10.4 The Power Poisson-Process Limit-Law

With Proposition 1 at our disposal, we are in position to establish the Poisson-process limit-law of Sect. 2.7. First, note that

$$\exp\left\{-\frac{1}{2}\left[\frac{\ln(x/y)-m(t)}{\sqrt{v(t)}}\right]^2\right\}$$
$$= \exp\left\{-\frac{1}{2v(t)}\left[\ln\left(\frac{x}{y}\right)\right]^2 + \frac{m(t)}{v(t)}\ln\left(\frac{x}{y}\right) - \frac{m(t)^2}{2v(t)}\right\} \qquad (2.45)$$
$$= \exp\left[-\frac{m(t)^2}{2v(t)}\right]\cdot\exp\left\{-\frac{1}{2v(t)}\left[\ln\left(\frac{x}{y}\right)\right]^2\right\}\cdot\left(\frac{x}{y}\right)^{m(t)/v(t)}.$$

Substituting Eq. (2.45) into Eq. (2.8) yields

$$f_t(x) = \frac{1}{c_t}g_t(x) \qquad (2.46)$$

($x > 0$), with

$$c_t = \sqrt{2\pi v(t)\exp\left[\frac{m(t)^2}{v(t)}\right]}, \qquad (2.47)$$

and with

$$g_t(x) = \int_0^\infty f_0(y) y^{-m(t)/v(t)} \exp\left\{-\frac{1}{2v(t)}\left[\ln\left(\frac{x}{y}\right)\right]^2\right\} dy \cdot x^{m(t)/v(t)-1}. \tag{2.48}$$

$(x > 0)$.

Equations (2.46)–(2.48) follow the setting of Proposition 1, with the domain $\mathcal{D}$ being the positive half-line $(x > 0)$. Equation (2.47), combined together with Eqs. (2.9) and (2.10), implies that Eq. (2.36) holds: $\lim_{t \to \infty} c_t = \infty$. Equation (2.48), combined together with Eqs. (2.9) and (2.10), implies that Eq. (2.37) holds with

$$g(x) = \int_0^\infty f_0(y) y^{-p} dy \cdot x^{p-1} = \mathbf{E}\left[X(0)^{-p}\right] \cdot x^{p-1}. \tag{2.49}$$

Also, in order to meet Eq. (2.38), Eq. (2.47) implies that Eq. (2.16) must hold. In turn, if Eq. (2.16) holds then we can apply Proposition 1. Doing so yields the Poisson-process limit ensemble $\mathcal{E}$ of Sect. 2.7—a Poisson process over the positive half-line with the intensity function of Eq. (2.17):

$$\lambda(x) = \omega g(x) = \omega \mathbf{E}[X(0)^{-p}] \cdot x^{p-1} \tag{2.50}$$

$(x > 0)$.

References

1. I. Eliazar, Physica A **512**, 1160 (2018)
2. C. Gardiner, *Handbook of Stochastic Methods* (Springer, New York, 2004)
3. N.G. Van Kampen, *Stochastic Processes in Physics and Chemistry*, 3rd edn. (North-Holland, Amsterdam, 2007)
4. Z. Schuss, *Theory and Applications of Stochastic Processes* (Springer, New York, 2009)
5. K. Ito, H.P. McKean, *Diffusion Processes and Their Sample Paths, Reprint of the*, 1974th edn. (Springer, Berlin, 1996)
6. I. Karatzas, S. Shreve, *Brownian Motion and Stochastic Calculus* (Springer, New York, 1991)
7. J.F. Le-Gall, *Brownian Motion, Martingales, and Stochastic Calculus* (Springer, New York, 2016)
8. A. Friedman, *Stochastic Differential Equations and Applications* (Dover, New York, 2006)
9. B. Oksendal, *Stochastic Differential Equations: An Introduction with Applications*, 6th edn. (Springer, New York, 2010)
10. L. Arnold, *Stochastic Differential Equations: Theory and Applications* (Dover, New York, 2011)
11. A. Hirsha, S.N. Neftci, *An Introduction to the Mathematics of Financial Derivatives* (Academic Press, London, 2013)
12. J.C. Hull, *Options, Futures, and Other Derivatives* (Pearson, London, 2017)
13. B. Dumas, E. Luciano, *The Economics of Continuous-time Finance* (MIT Press, Boston, 2017)
14. A.N. Borodin, P. Salminen, *Handbook of Brownian Motion - Facts and Formulae* (Birkhauser, Basel, 2015)
15. J.K. Patel, C.B. Read, *Handbook of the Normal Distribution* (Dekker, New York, 1996)

16. S. Redner, *A Guide To First-Passage Processes* (Cambridge University Press, Cambridge, 2001)
17. V. Seshadri, *The Inverse Gaussian Distribution* (Springer, New York, 1998)
18. I.A. Ibragimov, Y.A. Rozanov, *Gaussian Random Processes* (Springer, New York, 1978)
19. T. Hida, M. Hitsuda, *Gaussian Processes* (AMS, Providence, 2007)
20. D.R. Cox, *Statistics: An Appraisal*, ed. by H.A. David, H.T. David (Iowa State University Press, Ames, 1984), pp. 55–74
21. J. Beran, *Statistics for Long-memory Processes* (Chapman & Hall/CRC, London, 1994)
22. P. Doukhan, G. Oppenheim, M.S. Taqqu (eds.), *Theory and Applications of Long-Range Dependence* (Birkhauser, Boston, 2003)
23. G. Rangarajan, M. Ding (eds.), *Processes with Long-Range Correlations: Theory and Applications* (Springer, New York, 2003)
24. W.T. Coffey, Yu.P. Kalmykov, J.T. Waldron, *The Langevin Equation* (World Scientific, Singapore, 2012)
25. G.A. Pavliotis, *Stochastic Processes and Applications* (Springer, New York, 2014)
26. I. Nourdin, *Selected Aspects of Fractional Brownian Motion* (Springer, New York, 2012)
27. C. Berzin, A. Latour, J.R. Leon, *Inference on the Hurst Parameter and the Variance of Diffusions Driven by Fractional Brownian Motion* (Springer, New York, 2014)
28. J.F.C. Kingman, *Poisson Processes* (Oxford University Press, Oxford, 1993)
29. D.R. Cox, V. Isham, *Point Processes* (CRC, Boca Raton, 2000)
30. R.L. Streit, *Poisson Point Processes* (Springer, New York, 2010)
31. G. Peccati, M. Reitzner (eds.), *Stochastic Analysis for Poisson Point Processes* (Springer, New York, 2016)
32. G. Last, M. Penrose, *Lectures on the Poisson Process* (Cambridge University Press, Cambridge, 2017)
33. R. Brown, Phil. Mag. **4**, 161 (1828)
34. P.W. van der Pas, Scientiarum Historia **13**(1), 27 (1971)
35. P. Hanggi, F. Marchesoni, F. Nori, Annalen der Physik **14**, (2005)
36. A. Einstein, Ann. Physik **17**, 549 (1905)
37. M. von Smoluchowski, Ann. Physik **21**, 756 (1906)
38. J. Perrin, Comptes Rendus **146**, 967 (1908)
39. J. Perrin, Comptes rendus **147**, 475 (1908)
40. J. Perrin, Ann. Chim. Phys. **18**, 5 (1909)
41. N. Wiener, J. Math. Phys. **2**, 131 (1923)
42. K. Ito, Mem. Am. Math. Soc. **4**, 1 (1951)
43. M.D. Donsker, Mem. Am. Math. Soc. **6**, 1 (1951)
44. R. Paley, N. Wiener, A. Zigmund, Math Z. **37**, 647 (1933)
45. A. Dvoretski, P. Erdös, S. Kakutani, *Proceedings of the 4th Berkeley Symposium on Mathematical Statistics and Probability*, vol. 2 (1961), p. 103
46. P. Langevin, Comptes Rendus Acad. Sci. (Paris) **146**, 530 (1908)
47. G.E. Uhlenbeck, L.S. Ornstein, Phys. Rev. **36**, 823 (1930)
48. A.N. Kolmogorov, Dokl. Akad. Nauk SSSR **26**, 115 (1940)
49. A.M. Yaglom, Am. Math. Soc. Transl. **8**, 87 (1958)
50. B.B. Mandelbrot, Compt. Rend. **260**, 3274 (1965)
51. B.B. Mandelbrot, J.W. Van Ness, SIAM Rev. **10**, 422 (1968)
52. L. Bachelier, Ann. Sci. Ecole Norm. Sup. **17**, 21 (1900); L. Bachelier (translated by M. Davis and A. Etheridge), *Louis Bachelier's Theory of Speculation: The Origins of Modern Finance* (Princeton University Press, Princeton, 2006)
53. P.A. Samuelson, Indust. Manag. Rev. **6**, 13 (1965)
54. R.C. Merton, Bell J. Econ. Manag. Sci. **4**, 141 (1973)
55. F. Black, M. Scholes, J. Polit. Econ. **81**, 637 (1973)
56. F. Galton, Proc. R. Soc. (Lond.) **29**, 365 (1879)
57. R. Gibrat, *Les Inegalites Economiques* (Librairie du Recueil Sirey, Paris, 1931)
58. W.J. Adams, *The Life and Times of the Central Limit Theorem* (AMS, Providence, 2009)
59. H. Fischer, *A History of the Central Limit Theorem* (Springer, New York, 2010)

References

60. E. Mansfield, Am. Econ. Rev. **1023**, (1962)
61. J.M. Samuels, Rev. Econ. Stud. **105**, (1965)
62. J. Sutton, J. Econ. Lit. **35**, 40 (1997)
63. A.N. Kolmogorov, Dokl. Akad. Nauk SSSR. **31**, 99 (1941)
64. W. Shockley, Proc. IRE **45**, 279 (1957)
65. E.W. Montroll, M.F. Shlesinger, Proc. Natl. Acad. Sci. USA **79**, 3380 (1982)
66. E.W. Montroll, M.F. Shlesinger, J. Stat. Phys. **32**, 209 (1983)
67. M. Mitzenmacher, Internet Math. **1**, 226 (2004)
68. A.B. Downey, Comput. Commun. **28**, 790 (2005)
69. J. Aitchison, J.A.C. Brown, *The Lognormal Distribution with Special Reference to Its Uses in Econometrics* (Cambridge University Press, Cambridge, 1957)
70. E.L. Crow, K. Shimizu (eds.), *Lognormal Distributions: Theory and Applications* (Marcel Dekker, New York, 1988)
71. E. Limpert, W. Stahel, M. Abbt, BioScience **51**, 341 (2001)

Chapter 3
Setting the Stage

In Chap. 2, we established the emergence of the power Poisson process. In this chapter, we set the stage for the chapters to come—in which the power Poisson process shall be analyzed comprehensively. We present a terse review of the Poisson law, and then present a framework for the analysis of the power Poisson process.

3.1 The Poisson Law

The *Poisson law* is a principal feature of Poisson processes [1–5]. Indeed, for a Poisson process over a domain: the number of points (of the Poisson process) that reside in a given sub-domain is a *Poisson* random variable, with a mean that is the integral of the intensity function (of the Poisson process) over the sub-domain. In this section, we present a brief review of the Poisson law [6, 7], which is also termed the "*Law of Small Numbers*" [8, 9].

Consider a Poisson random variable K. This random variable admits non-negative integer values according to the following probabilities:

$$\Pr(K = k) = \exp(-\gamma) \frac{\gamma^k}{k!} \tag{3.1}$$

($k = 0, 1, 2, \ldots$), where γ is a positive parameter that characterizes the Poisson law. In particular, the probability that the Poisson random variable K admits the value zero is given by

$$\Pr(K = 0) = \exp(-\gamma) . \tag{3.2}$$

The mean and the variance of the Poisson random variable K are both given by its parameter:

$$\mathbf{E}[K] = \gamma \quad \& \quad \mathbf{Var}[K] = \gamma . \tag{3.3}$$

Hence, both the Poisson mean and the Poisson variance characterize the Poisson law. The Poisson mean and the Poisson variance imply that the standardized random variable that corresponds to the Poisson random variable K is given by

$$\hat{K} = \frac{K - \gamma}{\sqrt{\gamma}}. \tag{3.4}$$

Namely, the standardized random variable $\hat{K}$ has zero mean and unit variance: $\mathrm{E}[\hat{K}] = 0$ and $\mathbf{Var}[\hat{K}] = 1$.

In the limit $\gamma \to \infty$, the standardized random variable $\hat{K}$ converges, in law, to a limiting *standard normal* (*Gauss*) random variable Z, i.e., with zero mean and unit variance [10]; the derivation of this result is detailed in the Methods. Consequently, for $\gamma \gg 1$, the random variable K/γ admits the following asymptotic approximation in law:

$$\frac{K}{\gamma} \simeq 1 + \frac{1}{\sqrt{\gamma}} Z, \tag{3.5}$$

where Z is a standard normal (Gauss) random variable. Equation (3.5) manifests both a "*Law of Large Numbers*" and a "*Central Limit Theorem*" [11, 12] for the random variable K/γ. The "Law of Large Numbers" means that—for $\gamma \gg 1$—the random variable K/γ is approximated by its unit mean. The "Central Limit Theorem" means that—for $\gamma \gg 1$—the fluctuations of the random variable K/γ about its unit mean are of the order $1/\sqrt{\gamma}$, and that the fluctuations' statistics are normal (Gauss).

In general, the fluctuations of the random variable K/γ about its unit mean are dominated by *Chernoff bounds* [13, 14]. To describe these bounds, we set the function $\varphi(u) = 1 - u + u \ln(u)$ ($0 < u < \infty$), and note that this function is positive valued for non-unit inputs: $u \neq 1 \Rightarrow \varphi(u) > 0$. Then, the Chernoff bounds are as follows. For $0 < u < 1$ the bound is given by

$$\Pr\left(\frac{K}{\gamma} < u\right) \leq \exp[-\gamma \varphi(u)]. \tag{3.6}$$

And, for $1 < u < \infty$ the bound is given by

$$\Pr\left(\frac{K}{\gamma} > u\right) \leq \exp[-\gamma \varphi(u)]. \tag{3.7}$$

The derivations of the Chernoff bounds of Eqs. (3.6) and (3.7) are detailed in the Methods.

3.2 Framework

In Chap. 2, we established the emergence of the power Poisson process $\mathcal{E}$. This Poisson process is defined over the positive half-line, and is governed by the following power intensity function:

3.2 Framework

$$\lambda(x) = Cx^{p-1} \tag{3.8}$$

($x > 0$), where the coefficient C is a positive parameter, and where the power p is a real parameter. In the chapters to come, we shall explore the power Poisson process $\mathcal{E}$ in detail, and we shall do so using the framework described in this section.

In Chap. 2, we set off from a linear ODE; the temporal behavior of the linear ODE's solution is determined by the *sign* of the parameter r, the geometric rate of the linear ODE. Adding white noise, we moved on from the linear ODE to an SDE; the temporal behavior of the SDE's solution is determined by the *sign* of the "compound" parameter $\mu = r - \frac{1}{2}\nu$, which amalgamates the SDE's drift parameter r and volatility parameter ν. Then, replacing the underlying linear Brownian motion by a general Gaussian motion, we further moved from the SDE to a general geometric stochastic evolution; this evolution involved the parameter p, the limiting mean-to-variance ratio of the Gaussian motion.

The parameter p turned out to be the exponent of the power intensity function of Eq. (3.8). As the signs of the parameters r and μ determine, respectively, the temporal behaviors of the solutions of the aforementioned linear ODE and SDE—the *sign* of the parameter p determines the structural behavior of the power Poisson process $\mathcal{E}$. If the exponent is negative ($p < 0$) then the power intensity function of Eq. (3.8) is not integrable at the origin ($x \to 0$), and is integrable at infinity ($x \to \infty$). Conversely, if the exponent is positive ($p > 0$) then the power intensity function of Eq. (3.8) is integrable at the origin ($x \to 0$), and is not integrable at infinity ($x \to \infty$). This "integrability disparity" has profound implications on the structure of the power Poisson process $\mathcal{E}$, as we shall see in the chapters to come.

Due to the very different structural behaviors of the power Poisson process $\mathcal{E}$, as determined by the sign of the parameter p, we henceforth use the following framework:

- For positive p we denote by $\mathcal{E}_+$ the power Poisson process, set $\epsilon = p$ and $c = C/\epsilon$, and denote by

$$\lambda_+(x) = c\epsilon x^{\epsilon-1} \tag{3.9}$$

($x > 0$) its power intensity function.
- For negative p we denote by $\mathcal{E}_-$ the power Poisson process, set $\epsilon = -p$ and $c = C/\epsilon$, and denote by

$$\lambda_-(x) = c\epsilon x^{-\epsilon-1} \tag{3.10}$$

($x > 0$) its power intensity function.

The coefficient c and the exponent ϵ appearing in the power intensity functions of Eqs. (3.9) and (3.10) are both positive parameters. Various properties of the power Poisson process $\mathcal{E}_-$ were investigated in [15]. In this monograph, we shall present a far more in-depth exploration of the power Poisson process $\mathcal{E}_-$, as well as a counterpart exploration of the power Poisson process $\mathcal{E}_+$.

On the one hand, the power intensity function of Eq. (3.10) has a single shape: it is monotone decreasing ($\lambda'_-(x) < 0$) for all exponents $\epsilon > 0$. On the other hand, the power intensity function of Eq. (3.9) has three different shapes: (i) it is monotone decreasing ($\lambda'_+(x) < 0$) for exponents $\epsilon < 1$; (ii) it is flat ($\lambda'_+(x) = 0$) at the exponent value $\epsilon = 1$; and (iii) it is monotone increasing ($\lambda'_+(x) > 0$) for exponents $\epsilon > 1$. These different shapes induce different statistical behaviors, as we shall see in the chapters to come.

The flat intensity function $\lambda_+(x) = c$ ($x > 0$), which corresponds to the exponent value $\epsilon = 1$ in Eq. (3.9), characterizes the "standard" Poisson process. This Poisson process is widely applied in science and engineering as a statistical model for events that occur randomly in time [1–5]. Indeed, considering the underlying positive half-line to represent a time axis, the flat intensity function manifests a constant rate of occurrence: the probability that an event will take place during the infinitesimal time interval $(t, t + dt)$ is $c \cdot dt$ (and, conversely, the probability that no event will take place during this time interval is $1 - c \cdot dt$).

Evidently, the power Poisson processes $\mathcal{E}_+$ and $\mathcal{E}_-$ do not cover the case $p = 0$. In this case, the intensity function of Eq. (3.8) is *harmonic*: $\lambda(x) = C/x$ ($x > 0$). The harmonic intensity is integrable neither at the origin ($x \to 0$) nor at infinity ($x \to \infty$). The *harmonic Poisson process* —which is governed by the harmonic intensity $\lambda(x) = C/x$ —has unique properties and features. The harmonic Poisson process was studied in detail in [16]. Hence, as noted in the introduction, in this monograph we shall address the power Poisson processes $\mathcal{E}_+$ and $\mathcal{E}_-$, and shall not address the harmonic Poisson process.

The power Poisson processes, $\mathcal{E}_+$ and $\mathcal{E}_-$, are related to each other by the reciprocation of their points. Indeed, considering the reciprocal transformation $x \mapsto 1/x$, and using a general result from the theory of Poisson processes (the "displacement theorem" [1]), we arrive at the two following conclusions. (I) Applying the reciprocal transformation to each and every point of the power Poisson process $\mathcal{E}_+$ yields, in law, the power Poisson process $\mathcal{E}_-$. (II) Applying the reciprocal transformation to each and every point of the power Poisson process $\mathcal{E}_-$ yields, in law, the power Poisson process $\mathcal{E}_+$. The derivations of these conclusions are detailed in the Methods.

3.3 Methods

3.3.1 Normal Limit for Poisson Random Variables

The generating function of the Poisson random variable K of Eq. (3.1) is

$$\mathbf{E}\left[z^K\right] = \exp\left[\gamma(z-1)\right] \tag{3.11}$$

(z complex). Consequently, the characteristic function of the standardized random variable $\hat{K}$ of Eq. (3.4) is

3.3 Methods

$$\mathbf{E}\left[\exp\left(i\theta\hat{K}\right)\right] = \mathbf{E}\left[\exp\left(i\theta\frac{K-\gamma}{\sqrt{\gamma}}\right)\right]$$

$$= \mathbf{E}\left[\exp\left(i\frac{\theta}{\sqrt{\gamma}}K\right)\right]\exp\left(-i\theta\sqrt{\gamma}\right) \quad (3.12)$$

$$= \exp\left\{\gamma\left[\exp\left(i\frac{\theta}{\sqrt{\gamma}}\right) - 1\right]\right\}\exp\left(-i\theta\sqrt{\gamma}\right)$$

$(-\infty < \theta < \infty)$. In turn, for $\gamma \gg 1$ we have

$$\exp\left\{\gamma\left[\exp\left(i\frac{\theta}{\sqrt{\gamma}}\right) - 1\right]\right\}\exp\left(-i\theta\sqrt{\gamma}\right)$$

$$= \exp\left\{\gamma\left[i\frac{\theta}{\sqrt{\gamma}} + \frac{1}{2}\left(i\frac{\theta}{\sqrt{\gamma}}\right)^2 + O\left(\frac{1}{\gamma\sqrt{\gamma}}\right)\right] - i\theta\sqrt{\gamma}\right\} \quad (3.13)$$

$$= \exp\left[-\frac{1}{2}\theta^2 + O\left(\frac{1}{\sqrt{\gamma}}\right)\right]$$

$(-\infty < \theta < \infty)$. Thus, combining together Eqs. (3.12) and (3.13) we obtain the following limit:

$$\lim_{\gamma \to \infty} \mathbf{E}\left[\exp\left(i\theta\hat{K}\right)\right] = \exp\left(-\frac{1}{2}\theta^2\right) \quad (3.14)$$

$(-\infty < \theta < \infty)$. The limit term on the right-hand side of Eq. (3.14) is the characteristic function of a standard normal (Gauss) random variable Z. Hence, we obtain that: in the limit $\gamma \to \infty$ the standardized random variable $\hat{K}$ converges, in law, to a standard normal (Gauss) random variable Z.

3.3.2 Chernoff Bounds for Poisson Random Variables

Consider the Poisson random variable K of Eq. (3.1), and introduce the function

$$\varphi(u) = 1 - u + u\ln(u) \quad (3.15)$$

$(0 < u < \infty)$. Note that this function is positive valued for non-unit inputs: $u \neq 1 \Rightarrow \varphi(u) > 0$.

Fix $1 < u < \infty$, and set the function

$$\varphi_1(\theta) = 1 + u\theta - \exp(\theta) \quad (3.16)$$

$(\theta \geq 0)$. For any positive θ, Markov's inequality and Eqs. (3.11) and (3.16) imply that

$$\Pr(K > \gamma u)$$

$$= \Pr\left[\exp(\theta K) > \exp(\theta \gamma u)\right]$$

$$\leq \frac{\mathrm{E}[\exp(\theta K)]}{\exp(\theta \gamma u)} = \frac{\exp\{\gamma[\exp(\theta)-1]\}}{\exp(\theta \gamma u)} \qquad (3.17)$$

$$= \exp[-\gamma \cdot \varphi_1(\theta)] \ .$$

The function $\varphi_1(\theta)$ attains a global maximum at $\theta_1 = \ln(u)$, and Eqs. (3.15) and (3.16) imply that $\varphi_1(\theta_1) = \varphi(u)$. Hence, setting $\theta_1 = \ln(u)$ in Eq. (3.17) yields the Chernoff bound of Eq. (3.7):

$$\Pr\left(\frac{K}{\gamma} > u\right) \leq \exp[-\gamma \varphi(u)] \ , \qquad (3.18)$$

where $1 < u < \infty$.

Fix $0 < u < 1$, and set the function

$$\varphi_2(\theta) = 1 - u\theta - \exp(-\theta) \qquad (3.19)$$

($\theta \geq 0$). For any positive θ, Markov's inequality and Eqs. (3.11) and (3.19) imply that

$$\Pr(K < \gamma u)$$

$$= \Pr\left[\exp(-\theta K) > \exp(-\theta \gamma u)\right]$$

$$\leq \frac{\mathrm{E}[\exp(-\theta K)]}{\exp(-\theta \gamma u)} = \frac{\exp\{\gamma[\exp(-\theta)-1]\}}{\exp(-\theta \gamma u)} \qquad (3.20)$$

$$= \exp[-\gamma \cdot \varphi_2(\theta)] \ .$$

The function $\varphi_2(\theta)$ attains a global maximum at $\theta_2 = -\ln(u)$, and Eqs. (3.15) and (3.19) imply that $\varphi_2(\theta_2) = \varphi(u)$. Hence, setting $\theta_2 = -\ln(u)$ in Eq. (3.20) yields the Chernoff bound of Eq. (3.6):

$$\Pr\left(\frac{K}{\gamma} < u\right) \leq \exp[-\gamma \varphi(u)] \ , \qquad (3.21)$$

where $0 < u < 1$.

3.3.3 Reciprocation of $\mathcal{E}_+$ and $\mathcal{E}_-$

Consider a Poisson process $\mathcal{P}$ over a real domain $\mathcal{X}$, with intensity function $\lambda_\mathcal{P}(x)$ ($x \in \mathcal{X}$). Further consider a monotone function $y = f(x)$ that maps the real domain $\mathcal{X}$ to another real domain $\mathcal{Y}$. Applying the map $x \mapsto y = f(x)$ to each and every point of the Poisson process $\mathcal{P}$, we obtain the collection of points

$$\mathcal{Q} = \{f(x)\}_{x \in \mathcal{P}} . \tag{3.22}$$

The "displacement theorem" of the theory of Poisson processes [1] asserts that $\mathcal{Q}$ is a Poisson process over the real domain $\mathcal{Y}$, with intensity function

$$\lambda_\mathcal{Q}(y) = \frac{\lambda_\mathcal{P}\left[f^{-1}(y)\right]}{\left|f'\left[f^{-1}(y)\right]\right|} \tag{3.23}$$

($y \in \mathcal{Y}$).

Now, set the domain $\mathcal{X}$ to be the positive half-line, and set $f(x) = 1/x$ ($x > 0$). For this setting the domain $\mathcal{Y}$ is the positive half-line, and Eq. (3.23) yields

$$\lambda_\mathcal{Q}(y) = \lambda_\mathcal{P}\left(\frac{1}{y}\right)\frac{1}{y^2} \tag{3.24}$$

($y > 0$). Combining together Eq. (3.24) and Eqs. (3.9) and (3.10) we obtain that: (i) $\mathcal{P} = \mathcal{E}_+ \Rightarrow \lambda_\mathcal{P}(x) = \lambda_+(x) = c\epsilon x^{\epsilon-1} \Rightarrow \lambda_\mathcal{Q}(y) = c\epsilon y^{-\epsilon-1} = \lambda_-(y) \Rightarrow \mathcal{Q} = \mathcal{E}_-$ (in law); and (ii) $\mathcal{P} = \mathcal{E}_- \Rightarrow \lambda_\mathcal{P}(x) = \lambda_-(x) = c\epsilon x^{-\epsilon-1} \Rightarrow \lambda_\mathcal{Q}(y) = c\epsilon y^{\epsilon-1} = \lambda_+(y) \Rightarrow \mathcal{Q} = \mathcal{E}_+$ (in law). Namely, the power Poisson processes, $\mathcal{E}_+$ and $\mathcal{E}_-$, are related to each other by the reciprocation of their points.

References

1. J.F.C. Kingman, *Poisson Processes* (Oxford University Press, Oxford, 1993)
2. D.R. Cox, V. Isham, *Point Processes* (CRC, Boca Raton, 2000)
3. R.L. Streit, *Poisson Point Processes* (Springer, New York, 2010)
4. G. Peccati, M. Reitzner (eds.), *Stochastic Analysis for Poisson Point Processes* (Springer, New York, 2016)
5. G. Last, M. Penrose, *Lectures on the Poisson Process* (Cambridge University Press, Cambridge, 2017)
6. S.D. Poisson, *Recherches sur la probabilite des jugements en matiere criminelle et en matiere civile* (Bachelier, Paris, 1837)
7. F.A. Haight, *Handbook of the Poisson Distribution* (Wiley, New York, 1967)
8. L. von Bortkiewicz, *Das gesetz der kleinen zahlen* (Teubner, Leipzig, 1898)
9. M. Falk, J. Husler, R.D. Reiss, *Laws of Small Numbers: Extremes and Rare Events* (Birkhauser, Basel, 2010)
10. J.K. Patel, C.B. Read, *Handbook of the Normal Distribution* (Dekker, New York, 1996)

11. W. Feller, *An Introduction to Probability Theory and its Applications*, vol. 1, 2nd edn. (Wiley, New York, 1968)
12. W. Feller, *An Introduction to Probability Theory and its Applications*, vol. 2, 2nd edn. (Wiley, New York, 1971)
13. F. den Hollander, *Large Deviations* (AMS, Providence, 2008)
14. S.R.S. Varadhan, *Large Deviations* (AMS, Providence, 2016)
15. I. Eliazar, J. Klafter, J. Stat. Phys. **131**, 487 (2008)
16. I. Eliazar, Ann. Phys. **380**, 168 (2017)

Chapter 4
Threshold Analysis

In this chapter, we explore the statistical behavior of the power Poisson processes $\mathcal{E}_+$ and $\mathcal{E}_-$ below and above an arbitrary positive threshold level l. As we shall see, this threshold analysis will give rise to: the Pareto law and its inverse counterpart; and the Weibull law and its inverse counterpart, the Fréchet law.

4.1 Mean Behavior

Consider the power Poisson process $\mathcal{E}_+$. Its power intensity $\lambda_+(x) = c\epsilon x^{\epsilon-1}$ is not integrable at infinity ($x \to \infty$), and is integrable at the origin ($x \to 0$). The non-integrability at infinity implies that the number of points above the threshold level l is infinite with probability one. The integrability at the origin implies that the number of points below the threshold level l is finite with probability one.

More specifically, denote by $N_+(l)$ the number of points of the power Poisson process $\mathcal{E}_+$ that reside below the threshold level l. Then, the Poisson-process statistics imply that $N_+(l)$ is a *Poisson* random variable with mean

$$\mathbf{E}[N_+(l)] = \int_0^l \lambda_+(x)\,dx = cl^\epsilon. \tag{4.1}$$

As noted in Chap. 3, the Poisson law of the random variable $N_+(l)$ is characterized by its mean $\mathbf{E}[N_+(l)]$.

Consider the power Poisson process $\mathcal{E}_-$. Its power intensity $\lambda_-(x) = c\epsilon x^{-\epsilon-1}$ is not integrable at the origin ($x \to 0$) and is integrable at infinity ($x \to \infty$). The non-integrability at the origin implies that the number of points below the threshold level l is infinite with probability one. The integrability at infinity implies that the number of points above the threshold level l is finite with probability one.

More specifically, denote by $N_-(l)$ the number of points of the power Poisson process $\mathcal{E}_-$ that reside above the threshold level l. Then, the Poisson-process statistics imply that $N_-(l)$ is a *Poisson* random variable with mean

$$\mathbf{E}[N_-(l)] = \int_l^\infty \lambda_-(x)\,dx = cl^{-\epsilon}. \tag{4.2}$$

As noted in Chap. 3, the Poisson law of the random variable $N_-(l)$ is characterized by its mean $\mathbf{E}[N_-(l)]$.

Equations (4.1) and (4.2) manifest *power relations* between the threshold level l and the means $\mathbf{E}[N_+(l)]$ and $\mathbf{E}[N_-(l)]$. These power relations *characterize*, respectively, the power Poisson processes $\mathcal{E}_+$ and $\mathcal{E}_-$. Indeed, with regard to Eq. (4.1): $\mathbf{E}[N_+(l)] = cl^\epsilon$ ($l > 0$) holds *if and only if* the underlying intensity function is $\lambda_+(x) = c\epsilon x^{\epsilon-1}$ ($x > 0$). And, with regard to Eq. (4.2): $\mathbf{E}[N_-(l)] = cl^{-\epsilon}$ ($l > 0$) holds *if and only if* the underlying intensity function is $\lambda_-(x) = c\epsilon x^{-\epsilon-1}$ ($x > 0$).

On the one hand, the monotone decreasing power relation of Eq. (4.2) has a single shape: it is convex for all exponents $\epsilon > 0$. On the other hand, the monotone increasing power relation of Eq. (4.1) has three different shapes: concave for exponents $\epsilon < 1$; linear at the exponent value $\epsilon = 1$; and convex for exponents $\epsilon > 1$. The linear shape characterizes the "standard" Poisson process, which we discussed in Chap. 3.

4.2 Asymptotic Behavior

In Sect. 4.1, we saw that the number of points $N_+(l)$ of the power Poisson process $\mathcal{E}_+$ that reside below the threshold level l is a Poisson random variable with mean $\mathbf{E}[N_+(l)] = cl^\epsilon$. As this mean diverges in the limit $l \to \infty$, we can apply Eq. (3.5) of Chap. 3 to the Poisson random variable $N_+(l)$ and obtain the following asymptotic approximation in law:

$$\frac{N_+(l)}{cl^\epsilon} \simeq 1 + \frac{1}{\sqrt{cl^\epsilon}} Z \tag{4.3}$$

($l \gg 1$), where Z is a standard normal (Gauss) random variable.

Tracking the number of points $N_+(l)$ of the power Poisson process $\mathcal{E}_+$ that reside below the threshold level l, while increasing the level l from zero to infinity, we gradually move from a stochastic behavior to a deterministic behavior. For small l the behavior of the random variable $N_+(l)$ is stochastic, and it is governed by a Poisson law with mean cl^ϵ. For large l the behavior of the random variable $N_+(l)$ is essentially deterministic, and it is governed by the asymptotic approximation of Eq. (4.3).

In Sect. 4.1 we saw that the number of points $N_-(l)$ of the power Poisson process $\mathcal{E}_-$ that reside above the threshold level l is a Poisson random variable with mean $\mathbf{E}[N_-(l)] = cl^{-\epsilon}$. As this mean diverges in the limit $l \to 0$, we can apply Eq. (3.5) of Chap. 3 to the Poisson random variable $N_-(l)$ and obtain the following asymptotic approximation in law:

$$\frac{N_-(l)}{cl^{-\epsilon}} \simeq 1 + \frac{1}{\sqrt{cl^{-\epsilon}}} Z \tag{4.4}$$

4.2 Asymptotic Behavior

($l \ll 1$), where Z is a standard normal (Gauss) random variable.

Tracking the number of points $N_-(l)$ of the power Poisson process $\mathcal{E}_-$ that reside above the threshold level l, while decreasing the level l from infinity to zero, we gradually move from a stochastic behavior to a deterministic behavior. For large l the behavior of the random variable $N_-(l)$ is stochastic, and it is governed by the Poisson law with mean $cl^{-\epsilon}$. For small l the behavior of the random variable $N_-(l)$ is essentially deterministic, and it is governed by the asymptotic approximation of Eq. (4.4).

4.3 Log-Log Behavior

Applying the logarithmic function to both sides of the Eq. (4.3), and using Taylor's expansion, we obtain the following asymptotic approximation in law:

$$\ln\left[N_+(l)\right] \simeq \ln(c) + \epsilon \ln(l) + \frac{1}{\sqrt{cl^\epsilon}} Z \tag{4.5}$$

($l \gg 1$), where Z is a standard normal (Gauss) random variable.

In fact, for *all* threshold levels we have

$$\ln\left[N_+(l)\right] = \ln(c) + \epsilon \ln(l) + \Delta_+(l) \tag{4.6}$$

($l > 0$), where $\Delta_+(l)$ is an error term. Specifically, $\Delta_+(l)$ is a real-valued random variable whose fluctuations about the origin are dominated by Chernoff bounds that are given in Table 4.1. These bounds follow from the Chernoff bounds of Eqs. (3.6) and (3.7) of Chap. 3 via the following substitutions: $K = N_+(l)$ and $\gamma = cl^\epsilon$.

Applying the logarithmic function to both sides of Eq. (4.4), and using Taylor's expansion, we obtain the following asymptotic approximation in law:

$$\ln\left[N_-(l)\right] \simeq \ln(c) - \epsilon \ln(l) + \frac{1}{\sqrt{cl^{-\epsilon}}} Z \tag{4.7}$$

($l \ll 1$), where Z is a standard normal (Gauss) random variable.

In fact, for *all* threshold levels, we have

$$\ln\left[N_-(l)\right] = \ln(c) - \epsilon \ln(l) + \Delta_-(l) \tag{4.8}$$

($l > 0$), where $\Delta_-(l)$ is an error term. Specifically, $\Delta_-(l)$ is a real-valued random variable whose fluctuations about the origin are dominated by Chernoff bounds that are given in Table 4.1. These bounds follow from the Chernoff bounds of Eqs. (3.6) and (3.7) of Chap. 3 via the following substitutions: $K = N_-(l)$ and $\gamma = cl^{-\epsilon}$.

Table 4.1 Chernoff bounds for the error terms $\Delta_+(l)$ and $\Delta_-(l)$ of Eqs. (4.6) and (4.8), respectively. The function $\zeta(z)$ appearing in these bounds is given by $\zeta(z) = 1 + (z-1)\exp(z)$ ($-\infty < z < \infty$). This function is positive valued for all non-zero inputs: $z \neq 0 \Rightarrow \zeta(z) > 0$. The Chernoff bounds of the error terms $\Delta_+(l)$ and $\Delta_-(l)$ follow a super-exponential decay in the level variable l (as $l \to \infty$ for $\Delta_+(l)$, and as $l \to 0$ for $\Delta_-(l)$)

	$\mathcal{E}_+$	$\mathcal{E}_-$
$z < 0$	$\Pr[\Delta_+(l) < z] \leq$ $\exp[-cl^\epsilon \zeta(z)]$	$\Pr[\Delta_-(l) < z] \leq$ $\exp[-cl^{-\epsilon} \zeta(z)]$
$z > 0$	$\Pr[\Delta_+(l) > z] \leq$ $\exp[-cl^\epsilon \zeta(z)]$	$\Pr[\Delta_-(l) > z] \leq$ $\exp[-cl^{-\epsilon} \zeta(z)]$

Equations (4.6) and (4.8) manifest approximate *affine relations* between the *logarithm* $\ln(l)$ of the threshold level l and the *logarithms* $\ln[N_+(l)]$ and $\ln[N_-(l)]$. Namely, Eqs. (4.6) and (4.8) are, approximately, affine log-log plots of log-numbers versus log-levels.

4.4 Pareto Laws

Consider the points of the power Poisson process $\mathcal{E}_+$ that reside below the threshold level l. In Sect. 4.1, we saw that there are finitely many such points. A general result from the theory of Poisson processes implies that these finitely many points form a collection of IID random variables [1]. Moreover, the common probability density function of these IID random variables is given by the normalization of the intensity function $\lambda_+(x) = c\epsilon x^{\epsilon - 1}$ over the interval $(0, l)$.

Denote by $B_+(l)$ the position of a "representative point" of the power Poisson process $\mathcal{E}_+$ below the threshold level l. Then, the cumulative distribution function of the random variable $B_+(l)$ is given by:

$$\Pr[B_+(l) < x] = \frac{\int_0^x \lambda_+(x')\, dx'}{\int_0^l \lambda_+(x')\, dx'} = \left(\frac{x}{l}\right)^\epsilon \quad (4.9)$$

($0 < x < l$); the denominator appearing in the middle term of Eq. (4.9) is the normalization factor that corresponds to the interval $(0, l)$.

Consider the points of the power Poisson process $\mathcal{E}_-$ that reside above the threshold level l. In Sect. 4.1 we saw that there are finitely many such points. A general result from the theory of Poisson processes implies that these finitely many points form a collection of IID random variables [1]. Moreover, the common probability density function of these IID random variables is given by the normalization of the intensity function $\lambda_-(x) = c\epsilon x^{-\epsilon - 1}$ over the ray (l, ∞).

4.4 Pareto Laws

Denote by $A_-(l)$ the position of a "representative point" of the power Poisson process $\mathcal{E}_-$ above the threshold level l. Then, the tail distribution function of the random variable $A_-(l)$ is given by

$$\Pr[A_-(l) > x] = \frac{\int_x^\infty \lambda_-(x')\,dx'}{\int_l^\infty \lambda_-(x')\,dx'} = \left(\frac{l}{x}\right)^\epsilon \tag{4.10}$$

($l < x < \infty$); the denominator appearing in the middle term of Eq. (4.10) is the normalization factor that corresponds to the ray (l, ∞).

The tail distribution function of Eq. (4.10) characterizes the *Pareto law* [2, 3], which we discussed in the introduction. The cumulative distribution function of Eq. (4.9) characterizes the *inverse-Pareto law*. Note that the inverse-Pareto law with exponent $\epsilon = 1$ coincides with the *uniform law*: $\Pr[B_+(l) < x] = x/l$ ($0 < x < l$). We shall elaborate on this pair of Pareto laws in the next section.

4.5 Pareto Scaling

Consider the *inverse-Pareto* random variable $B_+(l)$, which was introduced in Sect. 4.4. This random variable manifests the position of a "representative point" of the power Poisson process $\mathcal{E}_+$ that resides below the threshold level l, and hence it takes values in the range $0 < B_+(l) < l$. Let us scale the random variable $B_+(l)$ with respect to the threshold level l. Namely, we shift from the random variable $B_+(l)$ to the scaled random variable $B_+(l)/l$.

Using Eq. (4.9) we obtain that the cumulative distribution function of the scaled random variable $B_+(l)/l$ is given by

$$\Pr\left[\frac{1}{l}B_+(l) < u\right] = \frac{\int_0^{lu} \lambda_+(x)\,dx}{\int_0^l \lambda_+(x)\,dx} = u^\epsilon \tag{4.11}$$

($0 < u < 1$). The right-hand side of Eq. (4.11) is invariant with respect to the threshold level l. Also, the right-hand side of Eq. (4.11) is the cumulative distribution function of the random variable $B_+(1)$. Hence, we obtain that the random variable $B_+(1)$ spans the continuum of the inverse-Pareto random variables $B_+(l)$ ($l > 0$) via the following *linear scaling*: $B_+(l) = l \cdot B_+(1)$, the equality being in law.

Equation (4.11) implies that the probability density function of the random variable $B_+(1)$ is given by: $\epsilon u^{\epsilon-1}$ ($0 < u < 1$). This density admits three different shapes, depending on the value of the exponent ϵ. (I) For exponent values $\epsilon < 1$ the density is monotone decreasing and unbounded, as it explodes near the origin. (II) At the exponent value $\epsilon = 1$, the density is constant. (III) For exponent values $\epsilon > 1$, the density is monotone increasing and bounded. Also, the mean of the random variable $B_+(1)$ is given by: $\mathbf{E}[B_+(1)] = \frac{\epsilon}{\epsilon+1}$.

Consider the *Pareto* random variable $A_-(l)$, which was introduced in Sect. 4.4. This random variable manifests the position of a "representative point" of the power Poisson process $\mathcal{E}_-$ that resides above the threshold level l, and hence it takes values in the range $l < A_-(l) < \infty$. Let us scale the random variable $A_-(l)$ with respect to the threshold level l. Namely, we shift from the random variable $A_-(l)$ to the scaled random variable $A_-(l)/l$.

Using Eq. (4.10), we obtain that the tail distribution function of the scaled random variable $A_-(l)/l$ is given by

$$\Pr\left[\frac{1}{l} A_-(l) > u\right] = \frac{\int_{lu}^\infty \lambda_-(x)\,dx}{\int_l^\infty \lambda_-(x)\,dx} = u^{-\epsilon} \tag{4.12}$$

($1 < u < \infty$). The right-hand side of Eq. (4.12) is invariant with respect to the threshold level l. Also, the right-hand side of Eq. (4.12) is the tail distribution function of the random variable $A_-(1)$. Hence, we obtain that the random variable $A_-(1)$ spans the continuum of the Pareto random variables $A_-(l)$ ($l > 0$) via the following *linear scaling*: $A_-(l) = l \cdot A_-(1)$, the equality being in law.

Equation (4.12) implies that the probability density function of the random variable $A_-(1)$ is given by: $\epsilon u^{-\epsilon-1}$ ($1 < u < \infty$). The shape of this density is monotone decreasing and bounded. There are two different scenarios for the mean of the random variable $A_-(1)$, depending on the value of the exponent ϵ. (I) For exponent values $\epsilon \leq 1$ the mean diverges, $\mathbf{E}[A_-(1)] = \infty$. (II) For exponent values $\epsilon > 1$ the mean is given by $\mathbf{E}[A_-(1)] = \frac{\epsilon}{\epsilon-1}$.

Equations (4.11) and (4.12) manifest the invariance of the scaled random variables $B_+(l)/l$ and $A_-(l)/l$ with respect to the threshold level l. These invariances *characterize*, respectively, the power Poisson processes $\mathcal{E}_+$ and $\mathcal{E}_-$. Indeed, with regard to Eq. (4.11): $\Pr[B_+(l)/l < u] = u^\epsilon$ ($l > 0$, $0 < u < 1$) holds *if and only if* the underlying intensity function is $\lambda_+(x) = c\epsilon x^{\epsilon-1}$ ($x > 0$). And, with regard to Eq. (4.12): $\Pr[A_-(l)/l > u] = u^{-\epsilon}$ ($l > 0$, $1 < u < \infty$) holds *if and only if* the underlying intensity function is $\lambda_-(x) = c\epsilon x^{-\epsilon-1}$ ($x > 0$).

The invariances of Eqs. (4.11) and (4.12) hold with their right-hand terms being *power functions*. This calls for the following question: can such invariances hold with right-hand terms that are *not* power functions? The answer to this question is negative. Indeed, with regard to Eq. (4.11): if $\Pr[B_+(l)/l < u] = \varphi_+(u)$ ($l > 0$, $0 < u < 1$) then the middle part of Eq. (4.11) implies that $\varphi_+(u_1 u_2) = \varphi_+(u_1)\varphi_+(u_2)$ ($0 < u_1, u_2 < 1$) – and hence $\varphi_+(u) = u^\epsilon$, where ϵ is a positive exponent. And, with regard to Eq. (4.12): if $\Pr[A_-(l)/l > u] = \varphi_-(u)$ ($l > 0$, $1 < u < \infty$) then the middle part of Eq. (4.12) implies that $\varphi_-(u_1 u_2) = \varphi_-(u_1)\varphi_-(u_2)$ ($1 < u_1, u_2 < \infty$) – and hence $\varphi_-(u) = u^{-\epsilon}$, where ϵ is a positive exponent.

4.6 Truncated Weibull Laws

Consider the points of the power Poisson process $\mathcal{E}_+$ that reside above the threshold level l. In Sect. 4.1, we saw that there are infinitely many such points. Among these infinitely many points, we set the spotlight on the *minimal point* and denote by $A_+(l)$ its position. Namely, $A_+(l)$ is the position of the point that is closest, from above, to the threshold level l.

The random variable $A_+(l)$ is greater than the level x (where $x > l$) if and only if the power Poisson process $\mathcal{E}_+$ has no points in the interval $(l, x]$. Namely: $A_+(l) > x \Leftrightarrow \mathcal{E}_+ \cap (l, x] = \emptyset$, where $\emptyset$ denotes the empty set. Consequently, the Poisson-process statistics and Eq. (3.2) of Chap. 3 imply that the tail distribution function of the random variable $A_+(l)$ is given by

$$\Pr\left[A_+(l) > x\right] = \exp\left[-\int_l^x \lambda_+(x')\,dx'\right] = \frac{\exp(cl^\epsilon)}{\exp(cx^\epsilon)} \quad (4.13)$$

$(l < x < \infty)$.

Consider the points of the power Poisson process $\mathcal{E}_-$ that reside below the threshold level l. In Sect. 4.1, we saw that there are infinitely many such points. Among these infinitely many points, we set the spotlight on the *maximal point* and denote by $B_-(l)$ its position. Namely, $B_-(l)$ is the position of the point that is closest, from below, to the threshold level l.

The random variable $B_-(l)$ is smaller than the level x (where $x < l$) if and only if the power Poisson process $\mathcal{E}_-$ has no points in the interval $[x, l)$. Namely: $B_-(l) < x \Leftrightarrow \mathcal{E}_- \cap [x, l) = \emptyset$, where, as above, $\emptyset$ denotes the empty set. Consequently, the Poisson-process statistics and Eq. (3.2) of Chap. 3 imply that the cumulative distribution function of the random variable $B_-(l)$ is given by

$$\Pr\left[B_-(l) < x\right] = \exp\left[-\int_x^l \lambda_-(x')\,dx'\right] = \frac{\exp(cl^{-\epsilon})}{\exp(cx^{-\epsilon})} \quad (4.14)$$

$(0 < x < l)$.

As we shall explain in the next section, the distribution functions of Eqs. (4.13) and (4.14) characterize, respectively, the *truncated Weibull laws*.

4.7 Weibull Laws

The random variable $A_+(l)$, which was introduced in Sect. 4.6, was defined with regard to a positive threshold level l. This definition can be extended to the zero threshold level $l = 0$. Doing so yields the random variable $A_+(0)$, which is the position of the *global minimal point* of the power Poisson process $\mathcal{E}_+$. Moreover, setting $l = 0$ in Eq. (4.13) yields the following tail distribution function:

$$\Pr\left[A_+(0) > x\right] = \exp\left[-\int_0^x \lambda_+(x')\,dx'\right] = \exp(-cx^\epsilon) \qquad (4.15)$$

($x > 0$).

Observe that, in terms of Eq. (4.15), the right-hand side of Eq. (4.13) coincides with the ratio $\Pr\left[A_+(0) > x\right] / \Pr\left[A_+(0) > l\right]$. This observation implies that the random variable $A_+(0)$ spans the continuum of the random variables $A_+(l)$ ($l > 0$) via the following *probabilistic conditioning*: the random variable $A_+(l)$ is equal in law to the random variable $A_+(0)$, given the information that $A_+(0)$ is greater than l. Hence, the random variable $A_+(l)$ is a "truncated version" of the random variable $A_+(0)$.

Equation (4.15) implies that the probability density function of the random variable $A_+(0)$ is given by: $c\epsilon \exp(-cx^\epsilon) x^{\epsilon-1}$ ($x > 0$). This density admits three different shapes, depending on the value of the exponent ϵ. (I) For exponent values $\epsilon < 1$, the density is monotone decreasing and unbounded, as it explodes near the origin. (II) At the exponent value $\epsilon = 1$, the density is monotone decreasing and bounded. (III) For exponent values $\epsilon > 1$, the density is unimodal. Also, the random variable $A_+(0)$ admits a finite mean.

The random variable $B_-(l)$, which was introduced in the Sect. 4.6, was defined with regard to a positive threshold level l. This definition can be extended to the infinite threshold level $l = \infty$. Doing so yields the random variable $B_-(\infty)$, which is the position of the *global maximal point* of the power Poisson process $\mathcal{E}_-$. Moreover, setting $l = \infty$ in Eq. (4.14) yields the following cumulative distribution function:

$$\Pr\left[B_-(\infty) < x\right] = \exp\left[-\int_x^\infty \lambda_-(x')\,dx'\right] = \exp(-cx^{-\epsilon}) \qquad (4.16)$$

($x > 0$).

Observe that, in terms of Eq. (4.16), the right-hand side of Eq. (4.14) coincides with the ratio $\Pr\left[B_-(\infty) < x\right] / \Pr\left[B_-(\infty) < l\right]$. This observation implies that the random variable $B_-(\infty)$ spans the continuum of the random variables $B_-(l)$ ($l > 0$) via the following *probabilistic conditioning*: the random variable $B_-(l)$ is equal in law to the random variable $B_-(\infty)$, given the information that $B_-(\infty)$ is smaller than l. Hence, the random variable $B_-(l)$ is a "truncated version" of the random variable $B_-(\infty)$.

Equation (4.16) implies that the probability density function of the random variable $B_-(\infty)$ is given by $c\epsilon \exp(-cx^{-\epsilon}) x^{-\epsilon-1}$ ($x > 0$). The shape of this density is unimodal. Also, there are two different scenarios for the mean of the random variable $B_-(\infty)$, depending on the value of the exponent ϵ. (I) For exponent values $\epsilon \leq 1$ the mean diverges. (II) For exponent values $\epsilon > 1$, the mean converges.

The tail distribution function of Eq. (4.15) characterizes the *Weibull law* [4, 5], which we discussed in the introduction. Note that the Weibull law with exponent $\epsilon = 1$ coincides with the *exponential law*: $\Pr\left[A_+(0) > x\right] = \exp(-cx)$ ($x > 0$). The cumulative distribution function of Eq. (4.16) characterizes the *Fréchet law*; this law is named in honor of the French mathematician Maurice Fréchet [6], and it is

4.7 Weibull Laws

also termed the *inverse-Weibull law*. The Weibull law and the Fréchet law constitute two of the three extreme-value laws [7, 8] that emerge universally from the Fisher–Tippett–Gnedenko theorem [9, 10].

The Weibull law and the Fréchet law *characterize*, respectively, the power Poisson processes $\mathcal{E}_+$ and $\mathcal{E}_-$. Indeed, with regard to Eq. (4.15): $\Pr\left[A_+(0) > x\right] = \exp\left(-cx^\epsilon\right)$ ($x > 0$) holds *if and only if* the underlying intensity function is $\lambda_+(x) = c\epsilon x^{\epsilon-1}$ ($x > 0$). And, with regard to Eq. (4.16): $\Pr\left[B_-(\infty) < x\right] = \exp\left(-cx^{-\epsilon}\right)$ ($x > 0$) holds *if and only if* the underlying intensity function is $\lambda_-(x) = c\epsilon x^{-\epsilon-1}$ ($x > 0$).

4.8 Outlook

We conclude with an outlook: an overview of the key facts and results that were presented along this chapter. As above, l is an arbitrary positive threshold level.

The power Poisson process $\mathcal{E}_+$ has finitely many points below the threshold level l, and has infinitely many points above the level l. The mean number of points below the threshold level l is given by the power function $\Lambda_+(l) = cl^\epsilon$, and this function characterizes the power Poisson process $\mathcal{E}_+$. The random variable $B_+(l)$ manifests the position of a "representative point" of the power Poisson process $\mathcal{E}_+$ that resides below the threshold level l; this random variable is inverse-Pareto, and the invariance of the scaled random variable $B_+(l)/l$ with respect to the threshold level l characterizes the power Poisson process $\mathcal{E}_+$. The random variable $A_+(l)$ manifests the position of the minimal point of the power Poisson process $\mathcal{E}_+$ above the level l, and the random variable $A_+(0)$ manifests the position of the global minimal point of the power Poisson process $\mathcal{E}_+$; the random variable $A_+(0)$ is Weibull, and it characterizes the power Poisson process $\mathcal{E}_+$. The connections between the power function $\Lambda_+(l) = cl^\epsilon$ and the random variables $B_+(l)$ and $A_+(l)$ are given in Table 4.2.

The power Poisson process $\mathcal{E}_-$ has infinitely many points below the threshold level l, and has finitely many points above the level l. The mean number of points above the threshold level l is given by the power function $\Lambda_-(l) = cl^{-\epsilon}$, and this function characterizes the power Poisson process $\mathcal{E}_-$. The random variable $A_-(l)$ manifests the position of a "representative point" of the power Poisson process $\mathcal{E}_-$ that resides above the threshold level l; this random variable is Pareto, and the invariance of the scaled random variable $A_-(l)/l$ with respect to the threshold level l characterizes the power Poisson process $\mathcal{E}_-$. The random variable $B_-(l)$ manifests the position of the maximal point of the power Poisson process $\mathcal{E}_-$ below the level l, and the random variable $B_-(\infty)$ manifests the position of the global maximal point of the power Poisson process $\mathcal{E}_-$; the random variable $B_-(\infty)$ is Fréchet (inverse-Weibull), and it characterizes the power Poisson process $\mathcal{E}_-$. The connections between the power function $\Lambda_-(l) = cl^{-\epsilon}$ and the random variables $B_-(l)$ and $A_-(l)$ are given in Table 4.2.

In Chap. 3, we described the reciprocal connection between the points of the power Poisson processes $\mathcal{E}_+$ and $\mathcal{E}_-$. This reciprocal connection is induced to the aforemen-

Table 4.2 The statistical behavior of the power Poisson processes, $\mathcal{E}_+$ and $\mathcal{E}_-$, below and above an arbitrary positive threshold level l. The $\mathcal{E}_+$ column summarizes the connection between: the mean number $\Lambda_+(l)$ of points that reside below the threshold l; the position $B_+(l)$ of a "representative point" below the threshold level l (with $0 < x < l$); and the position $A_+(l)$ of the point that is closest, from above, to the threshold level l (with $l < x < \infty$). The $\mathcal{E}_-$ column summarizes the connection between: the mean number $\Lambda_-(l)$ of points that reside above the threshold l; the position $B_-(l)$ of the point that is closest, from below, to the threshold level l (with $0 < x < l$); and the position $A_-(l)$ of a "representative point" above the threshold level l (with $l < x < \infty$)

	$\mathcal{E}_+$	$\mathcal{E}_-$
Below l	$\Pr\left[B_+(l) < x\right] = \frac{\Lambda_+(x)}{\Lambda_+(l)}$	$\Pr\left[B_-(l) < x\right] = \frac{\exp[\Lambda_-(l)]}{\exp[\Lambda_-(x)]}$
Above l	$\Pr\left[A_+(l) > x\right] = \frac{\exp[\Lambda_+(l)]}{\exp[\Lambda_+(x)]}$	$\Pr\left[A_-(l) > x\right] = \frac{\Lambda_-(x)}{\Lambda_-(l)}$

tioned quantities. First, the relationship between the power functions $\Lambda_+(l) = cl^\epsilon$ and $\Lambda_-(l) = cl^{-\epsilon}$ is given by: $\Lambda_+(l) = \Lambda_-(1/l)$ and $\Lambda_-(l) = \Lambda_+(1/l)$. Second, the relationship between the Pareto random variables $B_+(l)$ and $A_-(l)$ is given by $B_+(l) = 1/A_-(1/l)$ and $A_-(l) = 1/B_+(1/l)$, the equalities being in law. In particular, the relationship between the "keystone" Pareto random variables $B_+(1)$ and $A_-(1)$ is given by reciprocation: $B_+(1) = 1/A_-(1)$ and $A_-(1) = 1/B_+(1)$, the equalities being in law. Third, the relationship between the truncated Weibull random variables $A_+(l)$ and $B_-(l)$ is given by $A_+(l) = 1/B_-(1/l)$ and $B_-(l) = 1/A_+(1/l)$, the equalities being in law. Also, the relationship between the Weibull and Fréchet (inverse-Weibull) random variables $A_+(0)$ and $B_-(\infty)$ is given by reciprocation: $A_+(0) = 1/B_-(\infty)$ and $B_-(\infty) = 1/A_+(0)$, the equalities being in law.

References

1. J.F.C. Kingman, *Poisson Processes* (Oxford University Press, Oxford, 1993)
2. M. Hardy, Math. Intell. **32**, 38 (2010)
3. B.C. Arnold, *Pareto Distributions* (CRC Press, Boca Raton, 2015)
4. H. Rinne, *The Weibull Distribution: A Handbook* (CRC Press, Buca Raton, 2008)
5. J.I. McCool, *Using the Weibull Distribution: Reliability, Modeling and Inference* (Wiley, New York, 2012)
6. M. Fréchet, Ann. Soc. Polon. Math. Cracovie **6**, 93 (1927)
7. J. Galambos, *Asymptotic Theory of Extreme Order Statistics*, 2nd edn. (Krieger, Melbourne, 1987)
8. S. Kotz, S. Nadarajah, *Extreme Value Distributions* (Imperial College Press, London, 2000)
9. R.A. Fisher, L.H.C. Tippett, in *Proceedings of the Cambridge Philosophical Society*, vol. 24 (1928), pp. 180
10. B. Gnedenko, Ann. Math. **44**, 423 (1943) (translated and reprinted in: *Breakthroughs in Statistics I*, ed. by S. Kotz, N.L. Johnson, (Springer, New York, 1992), pp. 195–225

Chapter 5
Hazard Rates

In Chap. 4, we encountered the Weibull and Fréchet (inverse-Weibull) random variables associated with the power Poisson processes $\mathcal{E}_+$ and $\mathcal{E}_-$, respectively. In this short chapter, we use these Weibull random variables to present a *hazard-rate* perspective of the power Poisson processes $\mathcal{E}_+$ and $\mathcal{E}_-$. The notion of "hazard rate"—also termed "hazard function" and "failure rate"—is widely applied in reliability engineering [1, 2].

Consider an important event which we are anticipating, and which is expected to occur at some random time in the future. We start our observation at a time which we set to be our temporal origin, and denote by T the time at which the event will occur (relative to the temporal origin). The stochastic temporal nature of the event and the setting of the time axis imply that T is a positive-valued random variable.

Assume that the event did not occur before some positive time t. Then, what is the likelihood that the event will occur right at time t? The answer to this question is given by the following limit:

$$h_+(t) = \lim_{\Delta t \to 0} \frac{1}{\Delta t} \Pr(T < t + \Delta t | T \geq t). \tag{5.1}$$

The conditional probability appearing on the right-hand side of Eq. (5.1) is that the event will occur before time $t + \Delta t$, given the information that it did not occur before time t. We term the temporal function $h_+(t)$ ($t > 0$) the *forward hazard rate* of the random variable T.

There is a one-to-one correspondence between the forward hazard rate and the tail distribution function of the random variable T. Indeed, in terms of the forward hazard rate the tail distribution function is given by

$$\Pr(T > t) = \exp\left[-\int_0^t h_+(t') \, dt'\right] \tag{5.2}$$

($t > 0$). Equation (5.2) follows from Eq. (5.1) via a direct calculation.

Now, consider the Weibull random variable $A_+(0)$ of Chap. 4, the position of the global minimal point of the power Poisson process $\mathcal{E}_+$. As established in Chap. 4, the random variable $A_+(0)$ characterizes the power Poisson process $\mathcal{E}_+$, and its tail distribution function is given by

$$\Pr[A_+(0) > x] = \exp\left[-\int_0^x \lambda_+(x')\,dx'\right] \quad (5.3)$$

($x > 0$).

Observe that the tail distribution functions of Eqs. (5.2) and (5.3) are identical in form. This observation implies that the random variables $A_+(0)$ and T are equal in law if and only if the intensity function of the power Poisson process $\mathcal{E}_+$ coincides with the forward hazard rate:

$$\lambda_+(t) = c\epsilon t^{\epsilon-1} = h_+(t) \quad (5.4)$$

($t > 0$). Hence, we arrive at the following hazard-rate perspective: the intensity function of the power Poisson process $\mathcal{E}_+$ is the forward hazard rate of the position of its global minimal point, the Weibull random variable $A_+(0)$.

The hazard-rate perspective of the power Poisson process $\mathcal{E}_+$ implies three different behaviors, depending on the value of the exponent ϵ. (I) For exponent values $\epsilon < 1$ the forward hazard rate is monotone decreasing—manifesting an "anti-aging" behavior: the larger your age, the greater your chances to further live. (II) At the exponent value $\epsilon = 1$ the forward hazard rate is constant—manifesting a "no-memory" behavior: your chances to further live are independent of your age. (III) For exponent values $\epsilon > 1$ the forward hazard rate is a monotone increasing—manifesting an "aging" behavior: the larger your age, the smaller your chances to further live. Decreasing and increasing forward hazard rates play key roles in reliability engineering [1, 2].

Underlying the forward hazard rate was the implicit assumption that the time evolution is forward. But what if we reverse the temporal direction and move backward in time? In what follows, and in the context of the random variable T, we do just that: move backward in time. So, assume that the event did not occur after some positive time t. Then, the likelihood that the event will occur right at time t is given by the following limit:

$$h_-(t) = \lim_{\Delta t \to 0} \frac{1}{\Delta t} \Pr(T > t - \Delta t \mid T \leq t). \quad (5.5)$$

The conditional probability appearing on the right-hand side of Eq. (5.5) is that the event will occur after time $t - \Delta t$, given the information that it did not occur after time t. We term the temporal function $h_-(t)$ ($t > 0$) the *backward hazard rate* of the random variable T.

There is a one-to-one correspondence between the backward hazard rate and the cumulative distribution function of the random variable T. Indeed, in terms of the backward hazard rate the cumulative distribution function is given by:

$$\Pr(T < t) = \exp\left[-\int_t^\infty h_-(t')\,dt'\right] \tag{5.6}$$

($t > 0$). Equation (5.6) follows from Eq. (5.5) via a direct calculation.

Now, consider the Fréchet (inverse-Weibull) random variable $B_-(\infty)$ of Chap. 4, the position of the global maximal point of the power Poisson process $\mathcal{E}_-$. As established in Chap. 4, the random variable $B_-(\infty)$ characterizes the power Poisson process $\mathcal{E}_-$, and its cumulative distribution function is given by

$$\Pr\left[B_-(\infty) < x\right] = \exp\left[-\int_x^\infty \lambda_-(x')\,dx'\right] \tag{5.7}$$

($x > 0$).

Observe that the cumulative distribution functions of Eqs. (5.6) and (5.7) are identical in form. This observation implies that the random variables $B_-(\infty)$ and T are equal in law if and only if the intensity function of the power Poisson process $\mathcal{E}_-$ coincides with the backward hazard rate:

$$\lambda_-(t) = c\epsilon t^{-\epsilon-1} = h_-(t) \tag{5.8}$$

($t > 0$). Hence, we arrive at the following hazard-rate perspective: the intensity function of the power Poisson process $\mathcal{E}_-$ is the backward hazard rate of the position of its global maximal point, the Fréchet (inverse-Weibull) random variable $B_-(\infty)$.

References

1. E. Barlow, F. Proschan, *Mathematical Theory of Reliability* (SIAM, Philadelphia, 1996)
2. M. Finkelstein, *Failure Rate Modelling for Reliability and Risk* (Springer, New York, 2008)

Chapter 6
Lindy's Law

The Pareto laws emerged from the threshold analysis of Chap. 4. As these laws are scale invariant, they are characterized by a single parameter: the exponent ϵ. In this short chapter, we present altogether different characterizations of the Pareto laws: linear parameterizations based on conditional medians and on conditional means. These characterizations, which are based on [1], are generalizations of the Lindy law.

Lindy's was a New York deli at which comedians used to meet up after their shows [2]. Discussing successful and unsuccessful shows in these meet-ups, the comedians devised the following heuristic: consider a show that, to date, has been running for w weeks; then, the best estimate for the number of weeks the show will keep on running from now on is w. Namely, the best estimate for future performance is past performance. Termed *Lindy's law*, this heuristic can be applied to ideas and technologies—in order to predict how long will we still use them [3–6].

A statistical prediction is always with respect to an underlying metric. The most common metrics used in statistics are the Mean Absolute Deviation (MAD), and the Mean Square Deviation (MSD) [7–10]. From a geometric "Manhattan perspective" the MAD corresponds to the walking distance between two given junctions (the "L_1 norm" in math jargon), and the MSD corresponds to the aerial distance between the two junctions (the "L_2 norm" in math jargon). The *median* and the *mean* are, respectively, the best predictions with respect to the MAD and the MSD metrics.

As in Chap. 5, we consider an event which is due to occur at some random time in the future. Our time axis is the positive half-line, and we set T to be the time at which the event will occur. Thus, T is a positive-valued random variable, and we denote by $F(t) = \Pr(T < t) \; (t > 0)$ its cumulative distribution function, and by $\bar{F}(t) = \Pr(T > t) \; (t > 0)$ its tail distribution function.

Recalling Chap. 4, the random variable T is *Pareto* if

$$\bar{F}(t) = \left(\frac{l}{t}\right)^{\epsilon} \tag{6.1}$$

© Springer Nature Switzerland AG 2020
I. Eliazar, *Power Laws*, Understanding Complex Systems,
https://doi.org/10.1007/978-3-030-33235-8_6

($l < t < \infty$), where l is an arbitrary positive lower bound, and where ϵ is a positive exponent. Also, recalling Chap. 4, the random variable T is *inverse-Pareto* if

$$F(t) = \left(\frac{t}{l}\right)^{\epsilon} \tag{6.2}$$

($0 < t < l$), where l is an arbitrary positive upper bound, and where ϵ is a positive exponent.

Fix a positive time point t. Consider the scenario where we track the time axis up to time t, and observe that the event did not occur up to time t. Consequently, the duration we shall have to wait, from time t onward till observing the event, is $T - t$. With respect to the MAD metric, a generalized formulation of Lindy's law is

$$\mathbf{Med}\,[T - t \mid T > t] = \alpha \cdot t, \tag{6.3}$$

where α is a positive "Lindy parameter". Namely, the term appearing on the left-hand side of Eq. (6.3) is the conditional median of the waiting duration $T - t$ – given the information $\{T > t\}$. The right-hand side of Eq. (6.3) asserts that the conditional median is proportional to the time point t at which we are "standing".

The MAD formulation of Eq. (6.3) implies that

$$\frac{\bar{F}\,[(1+\alpha)\,t]}{\bar{F}\,(t)} = \frac{1}{2}\,; \tag{6.4}$$

the derivation of Eq. (6.4) is detailed in the Methods. Equation (6.4) has to hold for a varying time point t, and hence the tail distribution function appearing in it must be a monotone decreasing power function. In turn, a positive lower bound has to be introduced—and consequently Eq. (6.4) yields the Pareto tail distribution function of Eq. (6.1), with exponent

$$\epsilon = \frac{\ln(2)}{\ln(1+\alpha)}\,; \tag{6.5}$$

the derivation of Eq. (6.5) is detailed in the Methods.

The MAD formulation of Eq. (6.3) thus provides a characterization of the Pareto law via the 'Lindy parameter' α – the slope of the linear term appearing on the right-hand side of Eq. (6.3). The Lindy parameter $\alpha = 1$ is the borderline between two different Pareto regimes. (I) The finite-mean regime, $\mathbf{E}\,[T] < \infty$, which is characterized by the exponents $1 < \epsilon < \infty$ and by the slopes $0 < \alpha < 1$. (II) The infinite-mean regime, $\mathbf{E}\,[T] = \infty$, which is characterized by the exponents $0 < \epsilon \leq 1$ and by the slopes $1 \leq \alpha < \infty$.

Let's shift now from the MAD metric to the MSD metric. With respect to the MSD metric, a generalized formulation of Lindy's law is

$$\mathbf{E}\,[T - t \mid T > t] = \beta \cdot t, \tag{6.6}$$

where β is a positive "Lindy parameter". Similarly to Eq. (6.3), the term appearing on the left-hand side of Eq. (6.6) is the conditional mean of the waiting duration $T - t$ – given the information $\{T > t\}$. The right-hand side of Eq. (6.6) asserts that the conditional mean is proportional to the time point t at which we are "standing".

In Chap. 5 we introduced the forward hazard rate of the random variable T. Specifically, at the time point t, the forward hazard rate $h_+(t)$ has the following meaning: it is the likelihood that the event will occur right at time t, given the information that it did not occur up to time t. The MSD formulation of Eq. (6.6) implies that the forward hazard rate of the random variable T is harmonic:

$$h_+(t) = \underbrace{\frac{1+\beta}{\beta}}_{\epsilon} \cdot \frac{1}{t} ; \tag{6.7}$$

the derivation of Eq. (6.7) is detailed in the Methods. In turn, Eq. (6.7) yields the Pareto tail distribution function of Eq. (6.1), with exponent ϵ that is the coefficient appearing on the right-hand side of Eq. (6.7); see the Methods for the details.

The MSD formulation of Eq. (6.6) thus provides a characterization of the Pareto law via the "Lindy parameter" β—the slope of the linear term appearing on the right-hand side of Eq. (6.6). This characterization is restricted to the finite-mean regime, $\mathbf{E}[T] < \infty$, of the Pareto law: the exponent range $1 < \epsilon < \infty$. The Lindy parameter $\beta = 1$ is the borderline between two different Pareto regimes. (I) The finite-variance regime, $\mathbf{Var}[T] < \infty$, which is characterized by the exponents $2 < \epsilon < \infty$ and by the slopes $0 < \beta < 1$. (II) The infinite-variance regime, $\mathbf{Var}[T] = \infty$, which is characterized by the exponents $1 < \epsilon \leq 2$ and by the slopes $1 \leq \beta < \infty$.

So forth we implicitly assumed that the flow of time is forward. We now consider a time reversal, i.e., a backward flow of time. Specifically, having fixed a positive time point t as above, we consider the scenario where the event did not occur after time t. Consequently, with the information $\{T < t\}$ at hand, our goal is to predict when—within the time interval $(0, t)$—the event will occur.

With respect to the MAD metric, the counterpart formulation of Eq. (6.3) is

$$\mathbf{Med}[T \mid T < t] = \alpha \cdot t , \tag{6.8}$$

where α is a "Lindy parameter" that takes values in the unit interval ($0 < \alpha < 1$). Namely, the term appearing on the left-hand side of Eq. (6.8) is the conditional median of the random variable T—given the information $\{T < t\}$. The right-hand side of Eq. (6.8) asserts that the conditional median is proportional to the time point t at which we are "standing" (and from which we are looking backward in time).

The MAD formulation of Eq. (6.8) implies that

$$\frac{F(\alpha t)}{F(t)} = \frac{1}{2} ; \tag{6.9}$$

the derivation of Eq. (6.9) is detailed in the Methods. Equation (6.9) has to hold for a varying time point t, and hence the cumulative distribution function appearing in it must be a monotone increasing power function. In turn, a positive upper bound has to be introduced—and consequently Eq. (6.9) yields the inverse-Pareto cumulative distribution function of Eq. (6.2), with exponent

$$\epsilon = -\frac{\ln(2)}{\ln(\alpha)} ; \qquad (6.10)$$

the derivation of Eq. (6.10) is detailed in the Methods.

Let us shift now from the MAD metric to the MSD metric. With respect to the MSD metric, the counterpart formulation of Eq. (6.6) is

$$\mathbf{E}[T \mid T < t] = \beta \cdot t , \qquad (6.11)$$

where β is a "Lindy parameter" that takes values in the unit interval ($0 < \beta < 1$). Similar to Eq. (6.8), the term appearing on the left-hand side of Eq. (6.11) is the conditional mean of the random variable T—given the information $\{T < t\}$. The right-hand side of Eq. (6.11) asserts that the conditional mean is proportional to the time point t at which we are "standing" (and from which we are looking backward in time).

In Chap. 5, we introduced the backward hazard rate of the random variable T. Specifically, at the time point t, the backward hazard rate $h_-(t)$ has the following meaning: it is the likelihood that the event will occur right at time t, given the information that it did not occur after time t. The MSD formulation of Eq. (6.11) implies that the backward hazard rate of the random variable T is harmonic:

$$h_-(t) = \underbrace{\frac{\beta}{1-\beta}}_{\epsilon} \cdot \frac{1}{t} ; \qquad (6.12)$$

the derivation of Eq. (6.12) is detailed in the Methods. In turn, Eq. (6.12) yields the inverse-Pareto cumulative distribution function of Eq. (6.2), with exponent ϵ that is the coefficient appearing on the right-hand side of Eq. (6.12); see the Methods for the details.

The MAD formulation of Eq. (6.8) and the MSD formulation of Eq. (6.11) provide characterizations of the inverse-Pareto law via the "Lindy parameters" α and β— the slopes of the linear terms appearing, respectively, on the right-hand sides of Eqs. (6.8) and (6.11). The Lindy parameters $\alpha = \frac{1}{2}$ and $\beta = \frac{1}{2}$ correspond to the exponent $\epsilon = 1$—at which the inverse-Pareto law coincides with the uniform law.

Setting off from the MAD formulation of Eq. (6.3) and from the MSD formulation of Eq. (6.6), we arrived at the Pareto tail distribution function of Eq. (6.1); along the way, the lower bound of Eq. (6.1) emerged. Similarly, setting off from the MAD formulation of Eq. (6.8) and from the MSD formulation of Eq. (6.11), we arrived at the inverse-Pareto cumulative distribution function of Eq. (6.2); along the way, the

6 Lindy's Law

upper bound of Eq. (6.2) emerged. Evidently, the lower bound of Eq. (6.1) and the upper bound of Eq. (6.2) can assume any positive values. As we established in Chap. 4, the power Poisson processes $\mathcal{E}_-$ and $\mathcal{E}_+$—which are the respective bedrocks of the Pareto law and of the inverse-Pareto law— facilitate, simultaneously, all positive lower and upper bounds.

Methods

Equations (6.4) and (6.5)

The conditional tail distribution function of the waiting duration $T - t$— given the information $\{T > t\}$—is

$$\Pr(T - t > w \mid T > t) = \frac{\Pr(T > t + w)}{\Pr(T > t)} = \frac{\bar{F}(t+w)}{\bar{F}(t)} \tag{6.13}$$

($w > 0$). Denote by m the median of the conditional tail distribution function of Eq. (6.13). By definition

$$\frac{\bar{F}(t+m)}{\bar{F}(t)} = \frac{1}{2}. \tag{6.14}$$

Equation (6.14) implies that

$$m = \bar{F}^{-1}\left[\frac{1}{2}\bar{F}(t)\right] - t, \tag{6.15}$$

where $\bar{F}^{-1}(\cdot)$ is the inverse of the tail distribution function $\bar{F}(\cdot)$. Combining Eqs. (6.3) and (6.15) yields

$$\bar{F}^{-1}\left[\frac{1}{2}\bar{F}(t)\right] = (1+\alpha)t. \tag{6.16}$$

In turn, Eq. (6.16) implies that

$$\frac{1}{2} = \frac{\bar{F}[(1+\alpha)t]}{\bar{F}(t)}; \tag{6.17}$$

this proves Eq. (6.4).

As explained above, only the Pareto tail distribution function of Eq. (6.1) satisfies Eq. (6.17). Substituting the Pareto tail distribution function of Eq. (6.1) into Eq. (6.17) yields

$$\frac{1}{2} = (1+\alpha)^{-\epsilon}. \tag{6.18}$$

In turn, Eq. (6.18) yields Eq. (6.5).

Equation (6.7)

Equation (6.6) can be rewritten in the following form:

$$\mathbf{E}[T \mid T > t] = (1 + \beta) t . \qquad (6.19)$$

Denote by $f(x) = -\bar{F}'(x)$ the density function of the random variable T. The conditional density function of the random variable T—given the information $\{T > t\}$—is $f(x)/\bar{F}(t)$ $(x > t)$. In turn, the conditional mean of the random variable T—given the information $\{T > t\}$—is

$$\mathbf{E}[T \mid T > t] = \int_t^\infty x \cdot \frac{f(x)}{\bar{F}(t)} dx . \qquad (6.20)$$

Equations (6.19) and (6.20) imply that

$$\int_t^\infty x f(x) \, dx = (1 + \beta) t \bar{F}(t) . \qquad (6.21)$$

Differentiating Eq. (6.21) with respect to the variable t yields

$$-t f(t) = (1 + \beta) \left[\bar{F}(t) - t f(t) \right] . \qquad (6.22)$$

In turn, Eq. (6.22) implies that

$$\frac{f(t)}{\bar{F}(t)} = \frac{1 + \beta}{\beta} \frac{1}{t} . \qquad (6.23)$$

As the term appearing on the left-hand side of Eq. (6.23) is the forward hazard rate of the random variable T at the time point t—i.e., $h_+(t) = f(t)/\bar{F}(t)$—Eq. (6.23) implies Eq. (6.7).

We can rewrite Eq. (6.23) in the following form:

$$-\frac{d}{dt} \ln \left[\bar{F}(t) \right] = \frac{d}{dt} \left[\frac{1+\beta}{\beta} \ln(t) \right] . \qquad (6.24)$$

Integrating Eq. (6.24) yields

$$- \ln \left[\bar{F}(t) \right] = \frac{1+\beta}{\beta} \left[\ln(t) + r \right] , \qquad (6.25)$$

where r is a real integration constant. Setting $r = -\ln(l)$, Eq. (6.25) yields

$$\bar{F}(t) = \left(\frac{l}{t} \right)^\epsilon , \qquad (6.26)$$

with $\epsilon = (1 + \beta)/\beta$; this proves the Pareto law emanating from Eq. (6.6).

Equations (6.9) and (6.10)

The conditional cumulative distribution function of the random variable T —given the information $\{T < t\}$—is

$$\Pr(T < x \mid T < t) = \frac{\Pr(T < x)}{\Pr(T < t)} = \frac{F(x)}{F(t)} \quad (6.27)$$

$(0 < x < t)$. Denote by m the median of the conditional cumulative distribution function of Eq. (6.27). By definition

$$\frac{F(m)}{F(t)} = \frac{1}{2}. \quad (6.28)$$

Equation (6.28) implies that

$$m = F^{-1}\left[\frac{1}{2} F(t)\right], \quad (6.29)$$

where $F^{-1}(\cdot)$ is the inverse of the cumulative distribution function $F(\cdot)$. Combining Eqs. (6.8) and (6.27) yields

$$F^{-1}\left[\frac{1}{2} F(t)\right] = \alpha t. \quad (6.30)$$

In turn, Eq. (6.30) implies that

$$\frac{1}{2} = \frac{F(\alpha t)}{F(t)}; \quad (6.31)$$

this proves Eq. (6.9).

As explained above, only the inverse-Pareto cumulative distribution function of Eq. (6.2) satisfies Eq. (6.31). Substituting the inverse-Pareto cumulative distribution function of Eq. (6.2) into Eq. (6.31) yields

$$\frac{1}{2} = \alpha^{\epsilon}. \quad (6.32)$$

In turn, Eq. (6.32) yields Eq. (6.10).

Equation (6.12)

Denote by $f(x) = F'(x)$ the density function of the random variable T. The conditional density function of the random variable T—given the information $\{T < t\}$—is $f(x)/F(t)$ $(0 < x < t)$. In turn, the conditional mean of the random variable T—given the information $\{T < t\}$—is

$$\mathbf{E}[T \mid T < t] = \int_0^t x \cdot \frac{f(x)}{F(t)} dx. \quad (6.33)$$

Eqs. (6.11) and (6.33) imply that

$$\int_0^t x f(x)\, dx = \beta t F(t) \ . \tag{6.34}$$

Differentiating Eq. (6.34) with respect to the variable t yields

$$t f(t) = \beta [F(t) + t f(t)] \ . \tag{6.35}$$

In turn, Eq. (6.35) implies that

$$\frac{f(t)}{F(t)} = \frac{\beta}{1-\beta} \frac{1}{t} \ . \tag{6.36}$$

As the term appearing on the left-hand side of Eq. (6.36) is the backward hazard rate of the random variable T at the time point t—i.e., $h_-(t) = f(t)/F(t)$—Eq. (6.36) implies Eq. (6.12).

We can rewrite Eq. (6.36) in the following form:

$$\frac{d}{dt} \ln[F(t)] = \frac{d}{dt} \left[\frac{\beta}{1-\beta} \ln(t) \right] \ . \tag{6.37}$$

Integrating Eq. (6.24) yields

$$\ln[F(t)] = \frac{\beta}{1-\beta} [\ln(t) + r] \ , \tag{6.38}$$

where r is a real integration constant. Setting $r = -\ln(l)$, Eq. (6.38) yields

$$F(t) = \left(\frac{t}{l}\right)^\epsilon , \tag{6.39}$$

with $\epsilon = \beta/(1-\beta)$; this proves the inverse-Pareto law emanating from Eq. (6.11).

References

1. I. Eliazar, Physica A **486**, 797 (2017)
2. A. Goldman, New Republic **150**(24), 34 (1964)
3. B. Mandelbrot, *The Fractal Geometry of Nature* (Freeman, New York, 1982)
4. D. Sornette, D. Zajdenweber, Euro. Phys. J. B: Cond. Matt. Comp. Syst. **8**, 653 (1999)
5. N.N. Taleb, *Antifragile* (Random House, New York, 2012)
6. N.N. Taleb, *Skin in the Game* (Random House, New York, 2018)
7. A. Eddington, *Stellar Movements and the Structure of the Universe* (Macmillan, London, 1914)
8. R. Fisher, Mon. Notes R. Astron. Soc. **80**, 758 (1920)
9. S. Gorard, Brit. J. Educ. Stud. **53**, 417 (2005)
10. I. Eliazar, Phys. Rep. **649**, 1 (2016)

Chapter 7
Order Statistics

In this chapter we represent the power Poisson processes $\mathcal{E}_+$ and $\mathcal{E}_-$ as monotone sequences of order statistics, and explore the statistical behavior of these sequences. As we shall see, this order-statistics analysis will give rise to: the Zipf law and its inverse counterpart; the Weibull law and its inverse counterpart, the Fréchet law; and the Pareto law and its inverse counterpart.

7.1 Simulation

The most basic and well known Poisson process over the positive half-line, which is often termed "standard Poisson process", is characterized by a constant unit intensity. The points of the standard Poisson process can be represented via a sequence of IID exponential random variables, as we now describe.

Set $\{\xi_1, \xi_2, \xi_3, \ldots\}$ to be IID copies of an *exponential* random variable ξ with unit mean: $\Pr(\xi > t) = \exp(-t)$ $(t > 0)$. Then, the points of the standard Poisson process admit the following representation in law [1] : $\xi_1, \xi_1 + \xi_2, \xi_1 + \xi_2 + \xi_3, \ldots$. Namely, if we consider the points of the standard Poisson process to manifest a sequence of temporal events then: the first event occurs at time ξ_1, the second event occurs at time $\xi_1 + \xi_2$, the third event occurs at time $\xi_1 + \xi_2 + \xi_3$, etc.; in this temporal setting the random variable ξ_1 is the waiting-time between time zero and the first event, the random variable ξ_2 is the waiting-time between the first event and the second event, the random variable ξ_3 is the waiting-time between the second event and the third event, and so forth.

Consider the power transformation $x \mapsto (x/c)^{1/\epsilon}$, where the coefficient c and the exponent ϵ are positive parameters. A general result from the theory of Poisson processes (the "displacement theorem" [1]) implies that: Applying this power transformation to each and every point of the standard Poisson process yields, in law, the power Poisson process $\mathcal{E}_+$; see the Methods for the details. Hence, using the above representation for the points of the standard Poisson process, we obtain

© Springer Nature Switzerland AG 2020
I. Eliazar, *Power Laws*, Understanding Complex Systems,
https://doi.org/10.1007/978-3-030-33235-8_7

that the points of the power Poisson process $\mathcal{E}_+$ admit the following representation in law:

$$P_+(k) = \left(\frac{\xi_1 + \cdots + \xi_k}{c}\right)^{1/\epsilon} \tag{7.1}$$

($k = 1, 2, 3, \ldots$).

Equation (7.1) manifests the *order statistics* of the power Poisson process $\mathcal{E}_+$. These order statistics increase monotonically to infinity: $P_+(1) < P_+(2) < P_+(3) < \cdots \nearrow \infty$, the divergence to infinity holding with probability one. The first order statistic $P_+(1)$ coincides with the random variable $A_+(0)$ of Chap. 4—the position of the global minimal point of the power Poisson process $\mathcal{E}_+$.

Consider the power transformation $x \mapsto (c/x)^{1/\epsilon}$, where the coefficient c and the exponent ϵ are positive parameters. The aforementioned general result from the theory of Poisson processes (the "displacement theorem" [1]) implies that: Applying this power transformation to each and every point of the standard Poisson process yields, in law, the power Poisson process $\mathcal{E}_-$; see the Methods for the details. Hence, using the above representation for the points of the standard Poisson process, we obtain that the points of the power Poisson process $\mathcal{E}_-$ admit the following representation in law:

$$P_-(k) = \left(\frac{c}{\xi_1 + \cdots + \xi_k}\right)^{1/\epsilon} \tag{7.2}$$

($k = 1, 2, 3, \ldots$).

Equation (7.2) manifests the *order statistics* of the power Poisson process $\mathcal{E}_-$. These order statistics decrease monotonically to zero: $P_-(1) > P_-(2) > P_-(3) > \cdots \searrow 0$, the convergence to zero holding with probability one. The first order statistic $P_-(1)$ coincides with the random variable $B_-(\infty)$ of Chap. 4—the position of the global maximal point of the power Poisson process $\mathcal{E}_-$.

Equations (7.1) and (7.2) are highly efficient simulation algorithms for the order statistics of, respectively, the power Poisson processes $\mathcal{E}_+$ and $\mathcal{E}_-$. Note that the reciprocal connection between the points of the power Poisson processes $\mathcal{E}_+$ and $\mathcal{E}_-$, which was described in Chap. 3, is manifested by their respective order statistics. Indeed, it is evident from Eqs. (7.1) and (7.2) that: $P_+(k) = 1/P_-(k)$ and $P_-(k) = 1/P_+(k)$ ($k = 1, 2, 3, \ldots$), the equalities being in law.

7.2 Statistics

A calculation based on Eq. (7.1) implies that the probability density function of the kth order statistic of the power Poisson process $\mathcal{E}_+$ is given by:

$$\frac{d}{dx} \Pr\left[P_+(k) < x\right] = \frac{c^k \epsilon}{(k-1)!} \exp(-cx^\epsilon) x^{k\epsilon - 1} \tag{7.3}$$

7.2 Statistics

Table 7.1 The modes and the means of the order statistics $P_+(k)$ and $P_-(k)$. For the order statistic $P_+(k)$ the mode formula holds for $k\epsilon > 1$ (otherwise the mode vanishes). For the order statistic $P_-(k)$ the mean formula holds for $k\epsilon > 1$ (otherwise the mean diverges). In the mean formulae $\Gamma(u)$ ($u > 0$) denotes the Gamma function. The bottom row describes the common asymptotic behavior, in the limit $k \to \infty$, of the modes and the means; the asymptotic behavior of the means was calculated via Stirling's formula

	$P_+(k)$	$P_-(k)$
Mode	$\mathbf{M}[P_+(k)] = \left(\frac{k\epsilon-1}{c\epsilon}\right)^{1/\epsilon}$	$\mathbf{M}[P_-(k)] = \left(\frac{c\epsilon}{k\epsilon+1}\right)^{1/\epsilon}$
Mean	$\mathbf{E}[P_+(k)] = c^{-1/\epsilon} \frac{\Gamma(k+1/\epsilon)}{\Gamma(k)}$	$\mathbf{E}[P_-(k)] = c^{1/\epsilon} \frac{\Gamma(k-1/\epsilon)}{\Gamma(k)}$
$k \to \infty$	$\approx \left(\frac{k}{c}\right)^{1/\epsilon}$	$\approx \left(\frac{c}{k}\right)^{1/\epsilon}$

($x > 0$); the derivation of Eq. (7.3) is detailed in the Methods. For the first order statistic $P_+(1)$—which coincides with the random variable $A_+(0)$ of Chap. 4—Eq. (7.3) yields the Weibull probability density function which we encountered in Chap. 4: $c\epsilon \exp(-cx^\epsilon) x^{\epsilon-1}$ ($x > 0$).

The probability density function of Eq. (7.3) admits three different shapes, depending on the value of the product $k\epsilon$. (I) If $k\epsilon < 1$ then the density is monotone decreasing and unbounded, as it explodes near the origin. (II) If $k\epsilon = 1$ then the density is monotone decreasing and bounded. (III) If $k\epsilon > 1$ then the density is unimodal: it is monotone increasing in the range $x < \mathbf{M}[P_+(k)]$, it peaks at the mode $x = \mathbf{M}[P_+(k)]$, and it is monotone decreasing in the range $x > \mathbf{M}[P_+(k)]$. Also, the random variable $P_+(k)$ admits a finite mean. The values of the mode $\mathbf{M}[P_+(k)]$ and the mean $\mathbf{E}[P_+(k)]$ are given in Table 7.1.

A calculation based on Eq. (7.2) implies that the probability density function of the kth order statistic of the power Poisson process $\mathcal{E}_-$ is given by:

$$-\frac{d}{dx} \Pr[P_-(k) > x] = \frac{c^k \epsilon}{(k-1)!} \exp(-cx^{-\epsilon}) x^{-k\epsilon-1} \tag{7.4}$$

($x > 0$); the derivation of Eq. (7.4) is detailed in the Methods. For the first order statistic $P_-(1)$—which coincides with the random variable $B_-(\infty)$ of Chap. 4—Eq. (7.4) yields the Fréchet (inverse-Weibull) probability density function which we encountered in Chap. 4: $c\epsilon \exp(-cx^{-\epsilon}) x^{-\epsilon-1}$ ($x > 0$).

The shape of probability density function of Eq. (7.4) is unimodal: it is monotone increasing in the range $x < \mathbf{M}[P_-(k)]$, it peaks at the mode $x = \mathbf{M}[P_-(k)]$, and it is monotone decreasing in the range $x > \mathbf{M}[P_-(k)]$. Also, there are two different scenarios for the mean of the random variable $P_-(k)$, depending on the value of the product $k\epsilon$. (I) If $k\epsilon \leq 1$ then the mean diverges. (II) If $k\epsilon > 1$ then the mean converges. The values of the mode $\mathbf{M}[P_-(k)]$ and the mean $\mathbf{E}[P_-(k)]$ are given in Table 7.1.

7.3 Asymptotic Behavior

Equations (7.1) and (7.2) provide, respectively, representations of the order statistics of the power Poisson processes $\mathcal{E}_+$ and $\mathcal{E}_-$. These representations are based on a sequence $\{\xi_1, \xi_2, \xi_3, \ldots\}$ of IID exponential random variables with unit mean. For $k \gg 1$ the Central Limit Theorem implies the following asymptotic approximation, in law, for the averages of this sequence:

$$\frac{\xi_1 + \cdots + \xi_k}{k} \simeq 1 + \frac{1}{\sqrt{k}} Z, \tag{7.5}$$

where Z is a standard normal (Gauss) random variable Z, i.e. with zero mean and unit variance [2].

In Table 7.1 we saw that both the mode $\mathbf{M}\left[P_+(k)\right]$ and the mean $\mathbf{E}\left[P_+(k)\right]$ of the kth order statistic of the power Poisson process $\mathcal{E}_+$ are asymptotically equivalent, in the limit $k \to \infty$, to $(k/c)^{1/\epsilon}$. Combining together Eqs. (7.1) and (7.5), and using Taylor's expansion, we obtain the following asymptotic approximation in law:

$$\frac{P_+(k)}{(k/c)^{1/\epsilon}} \simeq 1 + \frac{1}{\epsilon\sqrt{k}} Z, \tag{7.6}$$

($k \gg 1$), where Z is as in Eq. (7.5).

Tracking the order statistics of the power Poisson process $\mathcal{E}_+$, while increasing the order-statistics rank k, we gradually move from a stochastic behavior to a deterministic behavior. For small k the behavior of the order statistic $P_+(k)$ is stochastic, and it is governed by the probability density function of Eq. (7.3). For large k the behavior of the order statistic $P_+(k)$ is essentially deterministic, and it is governed by the asymptotic approximation of Eq. (7.6).

In Table 7.1 we saw that both the mode $\mathbf{M}\left[P_-(k)\right]$ and the mean $\mathbf{E}\left[P_-(k)\right]$ of the kth order statistic of the power Poisson process $\mathcal{E}_-$ are asymptotically equivalent, in the limit $k \to \infty$, to $(c/k)^{1/\epsilon}$. Combining together Eqs. (7.2) and (7.5), and using Taylor's expansion, we obtain the following asymptotic approximation in law:

$$\frac{P_-(k)}{(c/k)^{1/\epsilon}} \simeq 1 + \frac{1}{\epsilon\sqrt{k}} Z, \tag{7.7}$$

($k \gg 1$), where Z is as in Eq. (7.5).

Tracking the order statistics of the power Poisson process $\mathcal{E}_-$, while increasing the order-statistics rank k, we gradually move from a stochastic behavior to a deterministic behavior. For small k the behavior of the order statistic $P_-(k)$ is stochastic, and it is governed by the probability density function of Eq. (7.4). For large k the behavior of the order statistic $P_-(k)$ is essentially deterministic, and it is governed by the asymptotic approximation of Eq. (7.7).

7.4 Zipf Laws

Applying the logarithmic function to both sides of Eq. (7.6), and using Taylor's expansion, we obtain the following asymptotic approximation in law:

$$\ln\left[P_+(k)\right] \simeq \frac{1}{\epsilon}\left[-\ln(c) + \ln(k) + \frac{1}{\sqrt{k}}Z\right] \quad (7.8)$$

($k \gg 1$), where Z is as in Eq. (7.5). In fact, for *all* the order statistics we have:

$$\ln\left[P_+(k)\right] = \frac{1}{\epsilon}\left[-\ln(c) + \ln(k) + \delta_k\right] \quad (7.9)$$

($k = 1, 2, 3, \ldots$), where δ_k is an error term on which we shall elaborate below.

Applying the logarithmic function to both sides of Eq. (7.7), and using Taylor's expansion, we obtain the following asymptotic approximation in law:

$$\ln\left[P_-(k)\right] \simeq \frac{1}{\epsilon}\left[\ln(c) - \ln(k) + \frac{1}{\sqrt{k}}Z\right] \quad (7.10)$$

($k \gg 1$), where Z is as in Eq. (7.5). In fact, for *all* the order statistics we have:

$$\ln\left[P_-(k)\right] = \frac{1}{\epsilon}\left[\ln(c) - \ln(k) - \delta_k\right] \quad (7.11)$$

($k = 1, 2, 3, \ldots$), where δ_k is an error term on which we shall elaborate now.

The error term δ_k appearing in Eqs. (7.9) and (7.11) is a real-valued random variable whose probability density function is unimodal, with mode $\mathbf{M}[\delta_k] = 0$. The fluctuations of the random variable δ_k about its mode are dominated by Chernoff bounds. To describe these bounds introduce the function $\zeta(z) = \exp(z) - 1 - z$ ($-\infty < z < \infty$), and note that this function is positive valued for non-zero inputs: $z \neq 0 \Rightarrow \zeta(z) > 0$. Then, the Chernoff bounds are as follows. (I) For $z < 0$ the bound is given by: $\Pr(\delta_k < z) \leq \exp[-k\zeta(z)]$. (II) For $z > 0$ the bound is given by: $\Pr(\delta_k > z) \leq \exp[-k\zeta(z)]$. The derivations of Eqs. (7.9) and (7.11), as well as the derivation of the statistics of the error term δ_k, are detailed in the Methods.

Equations (7.9) and (7.11) manifest approximate *affine relations* between the *logarithm* $\ln(k)$ of the order-statistics' rank and the *logarithms*, $\ln[P_+(k)]$ and $\ln[P_-(k)]$, of the order statistics. Namely, Eqs. (7.9) and (7.11) are, approximately, affine log-log plots of log-sizes versus log-ranks. The affine log-log plot of Eq. (7.11) – which was introduced and investigated in [3]—characterizes the *Zipf law* [4], which we discussed in the introduction. The affine log-log plot of Eq. (7.9) characterizes the *inverse-Zipf law*.

7.5 Ratios

Equation (7.1) implies that the order statistics of the power Poisson process $\mathcal{E}_+$ form a monotone increasing sequence. Hence, the $(k+1)$th order statistic $P_+(k+1)$ is larger than the kth order statistic $P_+(k)$. We now focus on the following quantitative question: how much larger is the $(k+1)$th order statistic than the kth order statistic?

To answer this question consider the ratio $P_+(k+1)/P_+(k)$. This ratio is a random variable, and a calculation using Eq. (7.1) implies that its tail distribution function is given by

$$\Pr\left[\frac{P_+(k+1)}{P_+(k)} > u\right] = u^{-k\epsilon} \tag{7.12}$$

$(1 < u < \infty)$; the derivation of Eq. (7.12) is detailed in the Methods. In turn, Eq. (7.12) implies that the cumulative distribution function of the inverse ratio $P_+(k)/P_+(k+1)$ is given by

$$\Pr\left[\frac{P_+(k)}{P_+(k+1)} < u\right] = u^{k\epsilon} \tag{7.13}$$

$(0 < u < 1)$.

Equation (7.2) implies that the order statistics of the power Poisson process $\mathcal{E}_-$ form a monotone decreasing sequence. Hence, the $(k+1)$th order statistic $P_+(k+1)$ is smaller than the kth order statistic $P_+(k)$. We now focus on the following quantitative question: how much smaller is the $(k+1)$th order statistic than the kth order statistic?

To answer this question consider the ratio $P_-(k+1)/P_-(k)$. This ratio is a random variable, and a calculation using Eq. (7.2) implies that its cumulative distribution function is given by

$$\Pr\left[\frac{P_-(k+1)}{P_-(k)} < u\right] = u^{k\epsilon} \tag{7.14}$$

$(0 < u < 1)$; the derivation of Eq. (7.14) is detailed in the Methods. In turn, Eq. (7.14) implies that the tail distribution function of the inverse ratio $P_-(k)/P_-(k+1)$ is given by

$$\Pr\left[\frac{P_-(k)}{P_-(k+1)} > u\right] = u^{-k\epsilon} \tag{7.15}$$

$(1 < u < \infty)$.

As noted in Chap. 4, the tail distribution functions of Eqs. (7.12) and (7.15) characterize the *Pareto law*, and the cumulative distribution functions of Eqs. (7.13) and (7.14) characterize the *inverse-Pareto law*. Note that the exponent of the Pareto laws of Eqs. (7.12)–(7.15) is the product $k\epsilon$—rather than the exponent ϵ, as in the Pareto laws of Chap. 4. The means of the ratios $P_+(k+1)/P_+(k)$ and $P_-(k+1)/P_-(k)$ are given in Table 7.2, where they are also compared to the ratios of the means of the corresponding order statistics.

7.6 Log-Ratios

Table 7.2 The means of the ratios $P_+(k+1)/P_+(k)$ and $P_-(k+1)/P_-(k)$, and the ratios of the means of the corresponding order statistics. For the ratio $P_+(k+1)/P_+(k)$ the mean formula holds for $k\epsilon > 1$ (otherwise the mean diverges). For the ratio of means $\mathbf{E}[P_-(k+1)]/\mathbf{E}[P_-(k)]$ the formula holds for $k\epsilon > 1$ (otherwise the ratio vanishes). The ratios of the means were calculated via Table 7.1

	$\mathcal{E}_+$	$\mathcal{E}_-$
Mean of ratios	$\mathbf{E}\left[\frac{P_+(k+1)}{P_+(k)}\right] = 1 + \frac{1}{k\epsilon - 1}$	$\mathbf{E}\left[\frac{P_-(k+1)}{P_-(k)}\right] = 1 - \frac{1}{k\epsilon + 1}$
Ratio of means	$\frac{\mathbf{E}[P_+(k+1)]}{\mathbf{E}[P_+(k)]} = 1 + \frac{1}{k\epsilon}$	$\frac{\mathbf{E}[P_-(k+1)]}{\mathbf{E}[P_-(k)]} = 1 - \frac{1}{k\epsilon}$

7.6 Log-Ratios

We now consider the logarithms of the ratios introduced in the previous section. Specifically: in the context of the power Poisson process $\mathcal{E}_+$ we focus on the log-ratio

$$L_+(k) = \ln\left[\frac{P_+(k+1)}{P_+(k)}\right] \tag{7.16}$$

($k = 1, 2, 3, \ldots$); and in the context of the power Poisson process $\mathcal{E}_-$ we focus on the log-ratio

$$L_-(k) = \ln\left[\frac{P_-(k)}{P_-(k+1)}\right] \tag{7.17}$$

($k = 1, 2, 3, \ldots$).

Equations (7.12) and (7.15) imply that the log-ratios $L_+(k)$ and $L_-(k)$ are equal in law, and that their common tail distribution function is

$$\Pr\left[L_+(k) > x\right] = \Pr\left[L_-(k) > x\right] = \exp(-k\epsilon x) \tag{7.18}$$

($x > 0; k = 1, 2, 3, \ldots$). The tail distribution function of Eq. (7.18) characterizes the *exponential law* with mean $1/(k\epsilon)$. Hence, there is a *harmonic relation* between the means of the log-ratios $L_+(k)$ and $L_-(k)$ and the rank k of the underlying order statistics.

The exponential law of Eq. (7.18) implies the following log-log representation for the log-ratios:

$$\ln[L_+(k)] = \ln[L_-(k)] = -\ln(\epsilon) - \ln(k) + \delta_1 \tag{7.19}$$

($k = 1, 2, 3, \ldots$), where the equalities in Eq. (7.19) are in law, and where δ_1 is the real-valued error term that we encountered in Sect. 7.4; the probability density function of this error term is unimodal, with mode $\mathbf{M}[\delta_1] = 0$.

Similarly to Eq. (7.11), Eq. (7.19) manifests the *Zipf law* [4]: an approximate affine log-log plot of log-sizes versus log-ranks. In the Zipf law of Eq. (7.11) the sizes under consideration were the order statistics $P_-(k)$ ($k = 1, 2, 3, \ldots$), and in the Zipf law of Eq. (7.19) the sizes under consideration are the log-ratios $L_+(k)$ and $L_-(k)$ ($k = 1, 2, 3, \ldots$). Note that these two Zipf laws are different. On the one hand, in the Zipf law of Eq. (7.11): the affine intercept is $\ln(c)/\epsilon$, the affine slope is $-1/\epsilon$, and the statistics of the error term δ_k depend on the rank k. On the other hand, in the Zipf law of Eq. (7.19): the affine intercept is $-\ln(\epsilon)$, the affine slope is -1, and the statistics of the error term δ_1 do not depend on the rank k.

7.7 Forward Motion

Consider a 'forward motion' on the order statistics of the power Poisson process $\mathcal{E}_+$: moving forward from $P_+(1)$ to $P_+(2)$, then from $P_+(2)$ to $P_+(3)$, then from $P_+(3)$ to $P_+(4)$, and so forth. This forward motion gives rise to the conditional law of the random variable $P_+(k+1)$, given the value of the random variable $P_+(k)$. A calculation based on Eq. (7.1) implies that this conditional law is given by the following conditional tail distribution function:

$$\Pr\left[P_+(k+1) > x | P_+(k) = l\right] = \frac{\exp(cl^\epsilon)}{\exp(cx^\epsilon)} \tag{7.20}$$

($l < x < \infty$); the derivation of Eq. (7.20) is detailed in the Methods.

The right-hand side of Eq. (7.20) coincides with the right-hand side of Eq. (4.13) of Chap. 4. Combining this coincidence with results from Chap. 4 we obtain that the three following random variables are equal in law. (I) The random variable $A_+(l)$, the position of the minimal point of the power Poisson process $\mathcal{E}_+$ above the threshold level l. (II) The first order statistic $P_+(1)$, given the information that it is greater than the threshold level l. (III) The order statistic $P_+(k+1)$, given the information that the order statistic $P_+(k)$ equals l.

Consider a 'forward motion' on the order statistics of the power Poisson process $\mathcal{E}_-$: moving forward from $P_-(1)$ to $P_-(2)$, then from $P_-(2)$ to $P_-(3)$, then from $P_-(3)$ to $P_-(4)$, and so forth. This forward motion gives rise to the conditional law of the random variable $P_-(k+1)$, given the value of the random variable $P_-(k)$. A calculation based on Eq. (7.2) implies that this conditional law is given by the following conditional cumulative distribution function:

$$\Pr\left[P_-(k+1) < x | P_-(k) = l\right] = \frac{\exp(cl^{-\epsilon})}{\exp(cx^{-\epsilon})} \tag{7.21}$$

($0 < x < l$); the derivation of Eq. (7.21) is detailed in the Methods.

The right-hand side of Eq. (7.21) coincides with the right-hand side of Eq. (4.14) of Chap. 4. Combining this coincidence with results from Chap. 4 we obtain that the

7.7 Forward Motion

three following random variables are equal in law. (I) The random variable $B_-(l)$, the position of the maximal point of the power Poisson process $\mathcal{E}_-$ below the threshold level l. (II) The first order statistic $P_-(1)$, given the information that it is smaller than the threshold level l. (III) The order statistic $P_-(k+1)$, given the information that the order statistic $P_-(k)$ equals l.

As noted in Chap. 4, the distribution functions of Eqs. (7.20) and (7.21) characterize the *truncated Weibull laws*. Moreover, in the limit $l \to 0$ the tail distribution function of Eq. (7.20) characterizes the *Weibull law*. And, in the limit $l \to \infty$ the tail distribution function of Eq. (7.21) characterizes the *Fréchet (inverse-Weibull) law*.

7.8 Backward Motion

Consider a 'backward motion' on the order statistics of the power Poisson process $\mathcal{E}_+$: moving counter-wise to the 'forward motion' described in the previous section. This backward motion gives rise to the conditional law of the random variable $P_+(k)$, given the value of the random variable $P_+(k+1)$. A calculation based on Eqs. (7.3) and (7.20), and that uses the Bayes formula, implies that this conditional law is given by the following conditional cumulative distribution function:

$$\Pr\left[P_+(k) < x | P_+(k+1) = l\right] = \left(\frac{x}{l}\right)^{k\epsilon} \tag{7.22}$$

$(0 < x < l)$; the derivation of Eq. (7.22) is detailed in the Methods.

In turn, Eq. (7.22) implies that the cumulative distribution function of the ratio $P_+(k)/P_+(k+1)$, given the value of the random variable $P_+(k+1)$, is:

$$\Pr\left[\frac{P_+(k)}{P_+(k+1)} < u | P_+(k+1) = l\right] = u^{k\epsilon} \tag{7.23}$$

$(0 < u < 1)$. The right-hand side of Eq. (7.23) coincides with the right-hand side of Eq. (7.13), and hence we obtain that: the ratio $P_+(k)/P_+(k+1)$ is *independent* of the random variable $P_+(k+1)$.

Consider a 'backward motion' on the order statistics of the power Poisson process $\mathcal{E}_-$: moving counter-wise to the 'forward motion' described in the previous section. This backward motion gives rise to the conditional law of the random variable $P_-(k)$, given the value of the random variable $P_-(k+1)$. A calculation based on Eqs. (7.4) and (7.21), and that uses the Bayes formula, implies that this conditional law is given by the following conditional tail distribution function:

$$\Pr\left[P_-(k) > x | P_-(k+1) = l\right] = \left(\frac{l}{x}\right)^{k\epsilon} \tag{7.24}$$

$(l < x < \infty)$; the derivation of Eq. (7.24) is detailed in the Methods.

In turn, Eq. (7.24) implies that the tail distribution function of the ratio $P_-(k)/P_-(k+1)$, given the value of the random variable $P_-(k+1)$, is:

$$\Pr\left[\frac{P_-(k)}{P_-(k+1)} > u | P_-(k+1) = l\right] = u^{-k\epsilon} \quad (7.25)$$

($1 < u < \infty$). The right-hand side of Eq. (7.25) coincides with the right-hand side of Eq. (7.15), and hence we obtain that: the ratio $P_-(k)/P_-(k+1)$ is *independent* of the random variable $P_-(k+1)$.

As noted in Chap. 4, the cumulative distribution functions of Eqs. (7.22) and (7.23) characterize the *inverse-Pareto law*, and the tail distribution functions of Eqs. (7.24) and (7.25) characterize the *Pareto law*. Note that the exponent of the Pareto laws of Eqs. (7.22)–(7.25) is the product $k\epsilon$—rather than the exponent ϵ, as in the Pareto laws of Chap. 4.

7.9 Outlook

We conclude with an outlook: an overview of the key facts and results that were presented along this chapter. As above, $k = 1, 2, 3, \ldots$ is the rank of the order statistics.

The points of the power Poisson process $\mathcal{E}_+$ form a monotone increasing sequence $P_+(1) < P_+(2) < P_+(3) < \cdots$; this sequence diverges to infinity with probability one. The elements of this sequence are positive-valued random variables that are termed the order statistics of the power Poisson process $\mathcal{E}_+$. The order statistic $P_+(k)$ admits the asymptotic power approximation $(k/c)^{1/\epsilon}$ (as $k \to \infty$). Also, on a log-log scale the order statistic $P_+(k)$ admits the affine log-log approximation $[\ln(k) - \ln(c)]/\epsilon$. The conditional law of the order statistic $P_+(k+1)$, given the information $P_+(k) = l$, is identical to the law of the random variable $A_+(l)$—the position of the minimal point of the power Poisson process $\mathcal{E}_+$ above the threshold level l. The conditional law of the order statistic $P_+(k)$, given the information $P_+(k+1) = l$, is inverse-Pareto with exponent $k\epsilon$. The law of the ratio $P_+(k)/P_+(k+1)$ is also inverse-Pareto with exponent $k\epsilon$, and on a log-log scale this ratio admits the affine log-log approximation $\ln(\epsilon) + \ln(k)$. The two affine log-log approximations—for the order statistic $P_+(k)$, and for the ratio $P_+(k)/P_+(k+1)$—manifest inverse-Zipf laws.

The points of the power Poisson process $\mathcal{E}_-$ form a monotone decreasing sequence $P_-(1) > P_-(2) > P_-(3) > \cdots$; this sequence converges to zero with probability one. The elements of this sequence are positive-valued random variables that are termed the order statistics of the power Poisson process $\mathcal{E}_-$. The order statistic $P_-(k)$ admits the asymptotic power approximation $(c/k)^{1/\epsilon}$ (as $k \to \infty$). Also, on a log-log scale the order statistic $P_-(k)$ admits the affine log-log approximation $[\ln(c) - \ln(k)]/\epsilon$. The conditional law of the order statistic $P_-(k+1)$, given the information $P_-(k) = l$, is identical to the law of the random variable $B_-(l)$—the position of the maximal point of the power Poisson process $\mathcal{E}_-$ below the thresh-

7.9 Outlook

Table 7.3 The forward and backward conditional distribution functions of the order statistics. The $\mathcal{E}_+$ column presents the conditional distribution functions of the random variables $P_+(k)$ and $P_+(k+1)$ in terms of the function $\Lambda_+(l)$, the mean number of points of $\mathcal{E}_+$ that reside below the threshold l. The $\mathcal{E}_-$ column presents the conditional distribution functions of the random variables $P_-(k)$ and $P_-(k+1)$ in terms of the function $\Lambda_-(l)$, the mean number of points of $\mathcal{E}_-$ that reside above the threshold l

	$\mathcal{E}_+$	$\mathcal{E}_-$
Forward	$\Pr\left[P_+(k+1) > x \mid P_+(k) = l\right]$ $= \exp\left[\Lambda_+(l)\right] / \exp\left[\Lambda_+(x)\right]$	$\Pr\left[P_-(k+1) < x \mid P_-(k) = l\right]$ $= \exp\left[\Lambda_-(l)\right] / \exp\left[\Lambda_-(x)\right]$
Backward	$\Pr\left[P_+(k) < x \mid P_+(k+1) = l\right]$ $= \left[\Lambda_+(x) / \Lambda_+(l)\right]^k$	$\Pr\left[P_-(k) > x \mid P_-(k+1) = l\right]$ $= \left[\Lambda_-(x) / \Lambda_-(l)\right]^k$

old level l. The conditional law of the order statistic $P_-(k)$, given the information $P_-(k+1) = l$, is Pareto with exponent $k\epsilon$. The law of the ratio $P_-(k)/P_-(k+1)$ is also Pareto with exponent $k\epsilon$, and on a log-log scale this ratio admits the affine log-log approximation $-\ln(\epsilon) - \ln(k)$. The two affine log-log approximations—for the order statistic $P_-(k)$, and for the ratio $P_-(k)/P_-(k+1)$ – manifest Zipf laws.

In Chap. 3 we described the reciprocal connection between the points of the power Poisson processes $\mathcal{E}_+$ and $\mathcal{E}_-$. In terms of the order statistics this reciprocal connection is straightforward: $P_+(k) = 1/P_-(k)$ and $P_-(k) = 1/P_+(k)$, the equalities being in law. As a comparison to Table 4.2 of Chap. 4, Table 7.3 of this chapter presents the forward and backward conditional distribution functions in terms of the power functions $\Lambda_+(l) = cl^\epsilon$ and $\Lambda_-(l) = cl^{-\epsilon}$—the mean number of points of the power Poisson process $\mathcal{E}_+$ that reside below the threshold l, and the mean number of points of the power Poisson process $\mathcal{E}_-$ that reside above the threshold l, respectively.

7.10 Methods

7.10.1 Simulation

Consider a Poisson process $\mathcal{P}$ over a real domain $\mathcal{X}$, with intensity function $\lambda_\mathcal{P}(x)$ ($x \in \mathcal{X}$). Further consider a monotone function $y = f(x)$ that maps the real domain $\mathcal{X}$ to another real domain $\mathcal{Y}$. Applying the map $x \mapsto y = f(x)$ to each and every point of the Poisson process $\mathcal{P}$, we obtain the collection of points

$$\mathcal{Q} = \{f(x)\}_{x \in \mathcal{P}} . \tag{7.26}$$

The "displacement theorem" of the theory of Poisson processes [1] asserts that $\mathcal{Q}$ is a Poisson process over the real domain $\mathcal{Y}$, with intensity function

$$\lambda_Q(y) = \frac{\lambda_P[f^{-1}(y)]}{|f'[f^{-1}(y)]|} \qquad (7.27)$$

($y \in \mathcal{Y}$).

Set the domain $\mathcal{X}$ to be the positive half-line, set $\lambda_P(x) = 1$ ($x > 0$), and: $f(x) = (x/c)^{1/p}$ ($x > 0$), where c is a positive constant, and where $p \neq 0$ is a real power. For this setting the domain $\mathcal{Y}$ is the positive half-line, and Eq. (7.27) yields

$$\lambda_Q(y) = c|p|y^{p-1} \qquad (7.28)$$

($y > 0$). Now, consider a given positive exponent ϵ. Then, combining together Eq. (7.28) with the definitions of the power Poisson processes $\mathcal{E}_+$ and $\mathcal{E}_-$ (set in Chap. 3) we obtain that: (i) $p = \epsilon \Rightarrow \lambda_Q(y) = c\epsilon y^{\epsilon-1} = \lambda_+(y) \Rightarrow Q = \mathcal{E}_+$ (in law); and (ii) $p = -\epsilon \Rightarrow \lambda_Q(y) = c\epsilon y^{-\epsilon-1} = \lambda_-(y) \Rightarrow Q = \mathcal{E}_-$ (in law).

7.10.2 Equations (7.3) and (7.4)

As in Sect. 7.1, set $\{\xi_1, \xi_2, \xi_3, \ldots\}$ to be IID copies of an exponential random variable ξ with unit mean: $\Pr(\xi > t) = \exp(-t)$ ($t > 0$). Also, denote by $\Sigma_k = \xi_1 + \cdots + \xi_k$ the sum of the first k IID copies. The sum Σ_k is an *Erlang* random variable [5]. Specifically, the law of the sum Σ_k is governed by the following *Erlang* probability density function [5]:

$$f_k(x) = \frac{d}{dx}\Pr(\Sigma_k < x) = \frac{1}{(k-1)!}\exp(-x)x^{k-1} \qquad (7.29)$$

($x > 0$).

In terms of the sum Σ_k Eq. (7.1) admits the following representation:

$$P_+(k) = \left(\frac{\Sigma_k}{c}\right)^{1/\epsilon}. \qquad (7.30)$$

Consequently, the cumulative distribution function of the random variable $P_+(k)$ is given by:

$$\Pr[P_+(k) < x] = \Pr\left[\left(\frac{\Sigma_k}{c}\right)^{1/\epsilon} < x\right] = \Pr(\Sigma_k < cx^\epsilon) \qquad (7.31)$$

($x > 0$). Differentiating both sides of Eq. (7.31), and using Eq. (7.29), implies that the probability density function of the random variable $P_+(k)$ is given by:

7.10 Methods

$$\frac{d}{dx}\Pr\left[P_+(k) < x\right] = \frac{d}{dx}\Pr\left(\Sigma_k < cx^\epsilon\right)$$

$$= f_k(cx^\epsilon) \cdot c\epsilon x^{\epsilon-1}$$

$$= \frac{1}{(k-1)!}\exp(-cx^\epsilon)(cx^\epsilon)^{k-1} \cdot c\epsilon x^{\epsilon-1} \quad (7.32)$$

$$= \frac{c^k \epsilon}{(k-1)!}\exp(-cx^\epsilon) x^{k\epsilon-1}$$

($x > 0$); this proves Eq. (7.3).

In terms of the sum Σ_k Eq. (7.2) admits the following representation:

$$P_-(k) = \left(\frac{c}{\Sigma_k}\right)^{1/\epsilon}. \quad (7.33)$$

Consequently, the tail distribution function of the random variable $P_-(k)$ is given by:

$$\Pr\left[P_-(k) > x\right] = \Pr\left[\left(\frac{c}{\Sigma_k}\right)^{1/\epsilon} > x\right] = \Pr\left(\Sigma_k < cx^{-\epsilon}\right) \quad (7.34)$$

($x > 0$). Differentiating both sides of Eq. (7.34), and using Eq. (7.29), implies that the probability density function of the random variable $P_-(k)$ is given by:

$$-\frac{d}{dx}\Pr\left[P_-(k) > x\right] = -\frac{d}{dx}\Pr\left(\Sigma_k < cx^{-\epsilon}\right)$$

$$= f_k(cx^{-\epsilon}) \cdot c\epsilon x^{-\epsilon-1}$$

$$= \frac{1}{(k-1)!}\exp\left(-cx^{-\epsilon}\right)(cx^{-\epsilon})^{k-1} \cdot c\epsilon x^{-\epsilon-1} \quad (7.35)$$

$$= \frac{c^k \epsilon}{(k-1)!}\exp\left(-cx^{-\epsilon}\right) x^{-k\epsilon-1}$$

($x > 0$); this proves Eq. (7.4).

7.10.3 Equations (7.9) and (7.11)

Set

$$\delta_k = \ln\left(\frac{\Sigma_k}{k}\right), \quad (7.36)$$

where $\Sigma_k = \xi_1 + \cdots + \xi_k$ is the sum of Eq. (7.29). Combining together Eqs. (7.30) and (7.36) yields:

$$\ln\left[P_+(k)\right] = \ln\left[\left(\frac{\Sigma_k}{c}\right)^{1/\epsilon}\right] = \frac{1}{\epsilon}\ln\left[\frac{k}{c}\frac{\Sigma_k}{k}\right]$$
$$= \frac{1}{\epsilon}\left[-\ln(c) + \ln(k) + \delta_k\right];\quad (7.37)$$

this proves Eq. (7.9). Combining together Eqs. (7.33) and (7.36) yields:

$$\ln\left[P_-(k)\right] = \ln\left[\left(\frac{c}{\Sigma_k}\right)^{1/\epsilon}\right] = \frac{1}{\epsilon}\ln\left[\frac{c}{k}\frac{k}{\Sigma_k}\right]$$
$$= \frac{1}{\epsilon}\left[\ln(c) - \ln(k) - \delta_k\right];\quad (7.38)$$

this proves Eq. (7.11).

Equation (7.36) implies that the cumulative distribution function of the random variable δ_k is given by:

$$\Pr(\delta_k < z) = \Pr\left[\ln\left(\frac{\Sigma_k}{k}\right) < z\right] = \Pr\left[\Sigma_k < k\exp(z)\right] \quad (7.39)$$

($-\infty < z < \infty$). Differentiating both sides of Eq. (7.39), and using Eq. (7.29), implies that the probability density function of the random variable δ_k is given by:

$$\begin{aligned}\frac{d}{dz}\Pr(\delta_k < z) &= \frac{d}{dz}\Pr\left[\Sigma_k < k\exp(z)\right] \\ &= f_k\left[k\exp(z)\right] \cdot k\exp(z) \\ &= \frac{1}{(k-1)!}\exp\left[-k\exp(z)\right]\left[k\exp(z)\right]^{k-1} \cdot k\exp(z) \\ &= \frac{k^k}{(k-1)!}\exp\left\{k\left[z - \exp(z)\right]\right\}\end{aligned} \quad (7.40)$$

($-\infty < z < \infty$). The shape of the probability density function of Eq. (7.40) is unimodal: it is monotone increasing along the negative half-line $z < 0$, it peaks at the origin $z = 0$, and it is monotone decreasing along the positive half-line $z > 0$. Hence, the mode of the random variable δ_k is the origin: $\mathbf{M}[\delta_k] = 0$.

Introduce the function

$$\zeta(z) = \exp(z) - 1 - z \quad (7.41)$$

($-\infty < z < \infty$), and recall that the Laplace transform of the Erlang random variable $\Sigma_k = \xi_1 + \cdots + \xi_k$ is given by [5]:

$$\mathbf{E}\left[\exp(\theta\Sigma_k)\right] = (1-\theta)^{-k} \quad (7.42)$$

($-\infty < \theta < 1$).

Consider $z < 0$, set $t = \exp(z)$, and note that $0 < t < 1$. Equation (7.39) implies that

$$\Pr(\delta_k < z) = \Pr(\Sigma_k < kt). \quad (7.43)$$

7.10 Methods

For $\theta < 0$ Eq. (7.43) implies that:

$$\Pr(\Sigma_k < kt) = \Pr\left[\exp(\theta \Sigma_k) > \exp(\theta kt)\right] \tag{7.44}$$

(using the Markov bound and Eq. (7.42))

$$\leq \frac{\mathbf{E}[\exp(\theta \Sigma_k)]}{\exp(\theta kt)} = \frac{(1-\theta)^{-k}}{\exp(\theta kt)}$$
$$= \exp\{-k[t\theta + \ln(1-\theta)]\}. \tag{7.45}$$

Eqs. (7.44)–(7.45) yield the inequality

$$\Pr(\Sigma_k < kt) \leq \exp\{-k[t\theta + \ln(1-\theta)]\} \tag{7.46}$$

($\theta < 0$). Consider the function $[t\theta + \ln(1-\theta)]$ ($\theta < 0$) appearing on the right-hand side of Eq. (7.46): the global maximum of this function, over the negative half-line, is attained at the point $\theta_* = 1 - 1/t$ (which is negative indeed, as $0 < t < 1$). In turn, setting $\theta = \theta_*$ in Eq. (7.46) yields

$$\Pr(\Sigma_k < kt) \leq \exp\{-k[t - 1 - \ln(t)]\}. \tag{7.47}$$

Combining together Eqs. (7.47) and (7.43), and using the notation of Eq. (7.41), we obtain that

$$\Pr(\delta_k < z) \leq \exp[-k \cdot \zeta(z)]. \tag{7.48}$$

Consider $z > 0$, set $t = \exp(z)$, and note that $t > 1$. Equation (7.39) implies that

$$\Pr(\delta_k > z) = \Pr(\Sigma_k > kt). \tag{7.49}$$

For $0 < \theta < 1$ Eq. (7.49) implies that:

$$\Pr(\Sigma_k > kt) = \Pr\left[\exp(\theta \Sigma_k) > \exp(\theta kt)\right]. \tag{7.50}$$

As in the transition from Eq. (7.44) to Eq. (7.46), Eq. (7.50) and the Markov bound yield the inequality

$$\Pr(\Sigma_k > kt) \leq \exp\{-k[t\theta + \ln(1-\theta)]\} \tag{7.51}$$

($0 < \theta < 1$). Consider the function $[t\theta + \ln(1-\theta)]$ ($0 < \theta < 1$) appearing on the right-hand side of Eq. (7.51): the global maximum of this function, over the unit interval, is attained at the point $\theta_* = 1 - 1/t$ (which is indeed within the unit interval, as $t > 1$). In turn, setting $\theta = \theta_*$ in Eq. (7.51) yields

$$\Pr(\Sigma_k > kt) \leq \exp\{-k[t - 1 - \ln(t)]\}. \tag{7.52}$$

Combining together Eqs. (7.52) and (7.49), and using the notation of Eq. (7.41), we obtain that
$$\Pr(\delta_k > z) \leq \exp[-k \cdot \zeta(z)]. \tag{7.53}$$

7.10.4 Equations (7.12) and (7.14)

Equation (7.1) and the sum $\Sigma_k = \xi_1 + \cdots + \xi_k$ of Eq. (7.29) imply that

$$\frac{P_+(k+1)}{P_+(k)} = \left(\frac{\xi_1 + \cdots + \xi_{k+1}}{c}\right)^{1/\epsilon} / \left(\frac{\xi_1 + \cdots + \xi_k}{c}\right)^{1/\epsilon}$$
$$= \left(\frac{\Sigma_k + \xi_{k+1}}{\Sigma_k}\right)^{1/\epsilon} = \left(1 + \frac{\xi_{k+1}}{\Sigma_k}\right)^{1/\epsilon}. \tag{7.54}$$

Using the fact that $\{\xi_1, \xi_2, \xi_3, \ldots\}$ are IID copies of a random variable ξ that is exponential with unit mean, and using Eq. (7.29), the tail distribution function of the ratio ξ_{k+1}/Σ_k is given by:

$$\Pr\left(\frac{\xi_{k+1}}{\Sigma_k} > t\right) = \int_0^\infty \Pr(\xi > tx) f_k(x) dx$$
$$= \int_0^\infty \exp(-tx) \frac{1}{(k-1)!} \exp(-x) x^{k-1} dx \tag{7.55}$$
$$= \int_0^\infty \frac{1}{(k-1)!} \exp[-(1+t)x] x^{k-1} dx = \frac{1}{(1+t)^k}$$

($t > 0$). Combining together Eqs. (7.54) and (7.55) we obtain that the tail distribution function of the ratio $P_+(k+1)/P_+(k)$ is given by:

$$\Pr\left(\frac{P_+(k+1)}{P_+(k)} > u\right) = \Pr\left[\left(1 + \frac{\xi_{k+1}}{\Sigma_k}\right)^{1/\epsilon} > u\right]$$
$$= \Pr\left(\frac{\xi_{k+1}}{\Sigma_k} > u^\epsilon - 1\right) = \frac{1}{[1+(u^\epsilon-1)]^k} = \frac{1}{u^{k\epsilon}} \tag{7.56}$$

($u > 1$); this proves Eq. (7.12).

Equation (7.2) and the sum $\Sigma_k = \xi_1 + \cdots + \xi_k$ of Eq. (7.29) imply that

$$\frac{P_-(k+1)}{P_-(k)} = \left(\frac{c}{\xi_1 + \cdots + \xi_{k+1}}\right)^{1/\epsilon} / \left(\frac{c}{\xi_1 + \cdots + \xi_k}\right)^{1/\epsilon}$$
$$\left(\frac{\Sigma_k}{\Sigma_k + \xi_{k+1}}\right)^{1/\epsilon} = \left(1 + \frac{\xi_{k+1}}{\Sigma_k}\right)^{-1/\epsilon}. \tag{7.57}$$

Combining together Eqs. (7.57) and (7.55) we obtain that the cumulative distribution function of the ratio $P_-(k+1)/P_-(k)$ is given by:

7.10 Methods

$$\Pr\left(\frac{P_-(k+1)}{P_-(k)} < u\right) = \Pr\left[\left(1 + \frac{\xi_{k+1}}{\Sigma_k}\right)^{-1/\epsilon} < u\right]$$ (7.58)

$$= \Pr\left(\frac{\xi_{k+1}}{\Sigma_k} > u^{-\epsilon} - 1\right) = \frac{1}{[1+(u^{-\epsilon}-1)]^k} = u^{k\epsilon}$$

$(0 < u < 1)$; this proves Eq. (7.14).

7.10.5 Equations (7.20) and (7.21)

Equation (7.30) implies that

$$\Sigma_k = c P_+(k)^\epsilon .$$ (7.59)

Substituting Eq. (7.59) into Eq. (7.54) yields

$$P_+(k+1) = \left(P_+(k)^\epsilon + \frac{\xi_{k+1}}{c}\right)^{1/\epsilon} .$$ (7.60)

Note that the random variable ξ_{k+1} is exponential with unit mean, and that it is independent of the random variable $P_+(k)$. Hence, from Eq. (7.60) we obtain that the conditional tail distribution function of the random variable $P_+(k+1)$ – given the information $P_+(k) = l$—is given by:

$$\Pr\left[P_+(k+1) > x | P_+(k) = l\right]$$
$$= \Pr\left[\left(l^\epsilon + \frac{\xi_{k+1}}{c}\right)^{1/\epsilon} > x\right]$$ (7.61)
$$= \Pr\left[\xi_{k+1} > c(x^\epsilon - l^\epsilon)\right]$$
$$= \exp\left[-c(x^\epsilon - l^\epsilon)\right] = \frac{\exp(cl^\epsilon)}{\exp(cx^\epsilon)}$$

$(l < x < \infty)$; this proves Eq. (7.20). In turn, the corresponding conditional probability density function is given by

$$\Phi_{k+1|k}(x;l) = \frac{\exp(cl^\epsilon)}{\exp(cx^\epsilon)} c\epsilon x^{\epsilon-1}$$ (7.62)

$(l < x < \infty)$.

Equation (7.33) implies that

$$\Sigma_k = c P_-(k)^{-\epsilon} .$$ (7.63)

Substituting Eq. (7.63) into Eq. (7.57) yields

$$P_-(k+1) = \left(P_-(k)^{-\epsilon} + \frac{\xi_{k+1}}{c}\right)^{-1/\epsilon}. \tag{7.64}$$

Note that the random variable ξ_{k+1} is exponential with unit mean, and that it is independent of the random variable $P_-(k)$. Hence, from Eq. (7.64) we obtain that the conditional cumulative distribution function of the random variable $P_-(k+1)$ – given the information $P_-(k) = l$—is given by:

$$\begin{aligned}
&\Pr\left[P_-(k+1) < x \mid P_-(k) = l\right] \\
&= \Pr\left[\left(l^{-\epsilon} + \frac{\xi_{k+1}}{c}\right)^{-1/\epsilon} < x\right] \\
&= \Pr\left[\xi_{k+1} > c\left(x^{-\epsilon} - l^{-\epsilon}\right)\right] \\
&= \exp\left[-c\left(x^{-\epsilon} - l^{-\epsilon}\right)\right] = \frac{\exp(cl^{-\epsilon})}{\exp(cx^{-\epsilon})}
\end{aligned} \tag{7.65}$$

($0 < x < l$); this proves Eq. (7.21). In turn, the corresponding conditional probability density function is given by

$$\Psi_{k+1|k}(x; l) = \frac{\exp(cl^{-\epsilon})}{\exp(cx^{-\epsilon})} c\epsilon x^{-\epsilon-1} \tag{7.66}$$

($0 < x < l$).

7.10.6 Equations (7.22) and (7.24)

Denote by $\Phi_k(x)$ ($x > 0$) the probability density function of Eq. (7.3). Also, denote by $\Phi_{k|k+1}(x; l)$ ($0 < x < l$) the conditional probability density function of the random variable $P_+(k)$, given the information $P_+(k+1) = l$. Using the Bayes formula, together with Eqs. (7.62) and (7.3), we obtain that

$$\begin{aligned}
\Phi_{k|k+1}(x; l) &= \Phi_{k+1|k}(l; x) \frac{\Phi_k(x)}{\Phi_{k+1}(l)} \\
&= \frac{\exp(cx^\epsilon)}{\exp(cl^\epsilon)} c\epsilon l^{\epsilon-1} \frac{\frac{c^k \epsilon}{(k-1)!} \exp(-cx^\epsilon) x^{k\epsilon-1}}{\frac{c^{k+1} \epsilon}{k!} \exp(-cl^\epsilon) l^{k\epsilon+\epsilon-1}} \\
&= k\epsilon \left(\frac{x}{l}\right)^{k\epsilon-1} \frac{1}{l}
\end{aligned} \tag{7.67}$$

7.10 Methods

($0 < x < l$). Integrating Eq. (7.67) with respect to the variable x we further obtain that the conditional cumulative distribution function of the random variable $P_+(k)$—given the information $P_+(k+1) = l$—is given by

$$\Pr\left[P_+(k) < x | P_+(k+1) = l\right] = \left(\frac{x}{l}\right)^{k\epsilon} \quad (7.68)$$

($0 < x < l$); this proves Eq. (7.22).

Denote by $\Psi_k(x)$ ($x > 0$) the probability density function of Eq. (7.4). Also, denote by $\Psi_{k|k+1}(x;l)$ ($l < x < \infty$) the conditional probability density function of the random variable $P_-(k)$, given the information $P_-(k+1) = l$. Using the Bayes formula, together with Eq. (7.66) and Eq. (7.4), we obtain that

$$\Psi_{k|k+1}(x;l) = \Psi_{k+1|k}(l;x) \frac{\Psi_k(x)}{\Psi_{k+1}(l)}$$

$$= \frac{\exp(cx^{-\epsilon})}{\exp(cl^{-\epsilon})} c\epsilon l^{-\epsilon-1} \frac{\frac{ck_\epsilon}{(k-1)!}\exp(-cx^{-\epsilon})x^{-k\epsilon-1}}{\frac{c^{k+1}\epsilon}{k!}\exp(-cl^{-\epsilon})l^{-k\epsilon-\epsilon-1}} \quad (7.69)$$

$$= k\epsilon \left(\frac{x}{l}\right)^{-k\epsilon-1} \frac{1}{l}$$

($l < x < \infty$). Integrating Eq. (7.69) with respect to the variable x we further obtain that the conditional tail distribution function of the random variable $P_-(k)$—given the information $P_-(k+1) = l$—is given by

$$\Pr\left[P_-(k) > x | P_-(k+1) = l\right] = \left(\frac{l}{x}\right)^{k\epsilon} \quad (7.70)$$

($l < x < \infty$); this proves Eq. (7.24).

References

1. J.F.C. Kingman, *Poisson Processes* (Oxford University Press, Oxford, 1993)
2. J.K. Patel, C.B. Read, *Handbook of the Normal Distribution* (Dekker, New York, 1996)
3. I. Eliazar, J. Phys. A: Math. Theor. **49**, 15LT01 (2016)
4. A. Saichev, D. Sornette, Y. Malevergne, *Theory of Zipf's Law and Beyond* (Springer, New York, 2009)
5. N.T. Thomopoulos, *Statistical Distributions* (Springer, New York, 2017)

Chapter 8
Exponent Estimation

Evidently, the principal parameter of the power Poisson processes, $\mathcal{E}_+$ and $\mathcal{E}_-$, is the *exponent* ϵ of their power intensities. In this short chapter, we address the estimation of the exponent ϵ via the observation of the first $m + 1$ order statistics of these Poisson processes. To that end, we take a deeper dive into the log-ratios of the order statistics (which were analyzed in Chap. 7).

Underlying the aforementioned log-ratios are logarithmic transformations: the transformation $x \mapsto \ln(x)$ in the case the power Poisson process $\mathcal{E}_+$, and the transformation $x \mapsto -\ln(x)$ in the case the power Poisson process $\mathcal{E}_-$. A result from the theory of Poisson processes (the "displacement theorem" [1]) implies that: Applying these logarithmic transformations, respectively, to each and every point of the power Poisson processes, $\mathcal{E}_+$ and $\mathcal{E}_-$, yields, in law, a Poisson process $\tilde{\mathcal{E}}$ over the real line with an *exponential intensity*:

$$\tilde{\lambda}(y) = c\epsilon \exp(\epsilon y) \tag{8.1}$$

($-\infty < y < \infty$); the derivation of Eq. (8.1) is detailed in the Methods. The exponential intensity of Eq. (8.1) manifests *accelerating change* [2–6].

The aforementioned logarithmic transformations imply that the points of the Poisson process $\tilde{\mathcal{E}}$ admit the following representation in law:

$$\tilde{P}(k) = \ln\left[P_+(k)\right] = -\ln\left[P_-(k)\right] \tag{8.2}$$

($k = 1, 2, 3, \ldots$), where $P_+(k)$ and $P_-(k)$ are the order statistics of the power Poisson processes $\mathcal{E}_+$ and $\mathcal{E}_-$, respectively. Equation (8.2) manifests the *order statistics* of the Poisson process $\tilde{\mathcal{E}}$. These order statistics increase monotonically to infinity: $\tilde{P}(1) < \tilde{P}(2) < \tilde{P}(3) < \cdots \nearrow \infty$, the divergence to infinity holding with probability one.

The consecutive gaps between these order statistics are given, in law, by the log-ratios that we encountered in Chap. 7:

$$\tilde{P}(k+1) - \tilde{P}(k) = \ln\left[\frac{P_+(k+1)}{P_+(k)}\right] = \ln\left[\frac{P_-(k)}{P_-(k+1)}\right] \qquad (8.3)$$

($k = 1, 2, 3, \ldots$). In Chap. 7, we established that the gap $\tilde{P}(k+1) - \tilde{P}(k)$ is an *exponential* random variable with mean $1/(k\epsilon)$. In the Methods, we further establish that the gaps $\tilde{P}(k+1) - \tilde{P}(k)$ ($k = 1, 2, 3, \ldots$) are *independent* random variables.

Thus, we obtain that

$$k \cdot \left[\tilde{P}(k+1) - \tilde{P}(k)\right] \qquad (8.4)$$

($k = 1, 2, 3, \ldots$) is a sequence of IID random variables that are governed by an *exponential law* with mean $1/\epsilon$. In turn, observing the first $m+1$ order statistics of the power Poisson processes $\mathcal{E}_+$ and $\mathcal{E}_-$ yields—via Eqs. (8.3) and (8.4)—a collection of m IID random variables that are exponential with mean $1/\epsilon$.

Considering these m exponential IID random variables, the *maximum likelihood estimator* (MLE) of their mean $1/\epsilon$ is their *empirical average* [7, 8]:

$$A(m) = \frac{1}{m} \sum_{k=1}^{m} k \cdot \left[\tilde{P}(k+1) - \tilde{P}(k)\right]$$

$$= \frac{1}{m} \sum_{k=1}^{m} \left[\tilde{P}(m+1) - \tilde{P}(k)\right] \qquad (8.5)$$

$$= \tilde{P}(m+1) - \frac{1}{m} \sum_{k=1}^{m} \tilde{P}(k) ;$$

the transition from the first line to the second line of Eq. (8.5) is attained by straightforward algebra. In turn, the MLE of the parameter ϵ is the *reciprocal* $1/A(m)$ of the empirical average [7, 8]. The key statistical features of these MLEs are detailed in Table 8.1.

Combining together Eqs. (8.5) and (8.2), we arrive at the following pair of results. In terms of the first $m+1$ order statistics of the power Poisson process $\mathcal{E}_+$, the empirical average is given by

$$A(m) = \ln[P_+(m+1)] - \frac{1}{m} \sum_{k=1}^{m} \ln[P_+(k)] . \qquad (8.6)$$

In terms of the first $m+1$ order statistics of the power Poisson process $\mathcal{E}_-$, the empirical average is given by

$$A(m) = -\ln[P_-(m+1)] + \frac{1}{m} \sum_{k=1}^{m} \ln[P_-(k)] . \qquad (8.7)$$

Eqs. (8.6) and (8.7) are efficient formulae for the computation of the empirical average $A(m)$.

8 Exponent Estimation

Table 8.1 The key statistical features of the MLEs of the parameters $1/\epsilon$ and ϵ. The term $A(m)$ is the empirical average given by Eq. (8.5) (in terms of the first $m+1$ order statistics of the Poisson process $\tilde{\mathcal{E}}$), and by Eqs. (8.6) and (8.7) (in terms of the first $m+1$ order statistics of the power Poisson processes $\mathcal{E}_+$ and $\mathcal{E}_-$, respectively). The formulae for the mode of the MLE of $1/\epsilon$ and for the mean of the MLE of ϵ hold for $m+1 > 2$ observations. The formula for the variance of the MLE of ϵ holds for $m+1 > 3$ observations

	$1/\epsilon$	ϵ
MLE	$A(m)$	$1/A(m)$
Mode	$\mathbf{M}[A(m)] = \frac{m-1}{m} \frac{1}{\epsilon}$	$\mathbf{M}\left[\frac{1}{A(m)}\right] = \frac{m}{m+1}\epsilon$
Mean	$\mathbf{E}[A(m)] = \frac{1}{\epsilon}$	$\mathbf{E}\left[\frac{1}{A(m)}\right] = \frac{m}{m-1}\epsilon$
Variance	$\mathbf{Var}[A(m)] = \frac{1}{m}\left(\frac{1}{\epsilon}\right)^2$	$\mathbf{Var}\left[\frac{1}{A(m)}\right] = \frac{1}{m-2}\left(\frac{m}{m-1}\right)^2 \epsilon^2$

In conclusion, an estimation algorithm for the power Poisson processes $\mathcal{E}_+$ and $\mathcal{E}_-$ is summarized by the three following steps. Step I: observe the first $m+1$ order statistics of the power Poisson processes, and compute the empirical average $A(m)$ via Eqs. (8.6) and (8.7). Step II: to estimate the parameter ϵ —the exponent of the power intensities of the power Poisson processes— use the MLE $1/A(m)$. Step III: to estimate the parameter $1/\epsilon$—the slope of the Zipf laws of the order statistics of the power Poisson processes—use the MLE $A(m)$.

Methods

Equation (8.1)

Consider a Poisson process $\mathcal{P}$ over a real domain $\mathcal{X}$, with intensity function $\lambda_\mathcal{P}(x)$ ($x \in \mathcal{X}$). Further consider a monotone function $y = f(x)$ that maps the real domain $\mathcal{X}$ to another real domain $\mathcal{Y}$. Applying the map $x \mapsto y = f(x)$ to each and every point of the Poisson process $\mathcal{P}$, we obtain the collection of points

$$\mathcal{Q} = \{f(x)\}_{x \in \mathcal{P}}. \tag{8.8}$$

The "displacement theorem" of the theory of Poisson processes [1] asserts that $\mathcal{Q}$ is a Poisson process over the real domain $\mathcal{Y}$, with intensity function

$$\lambda_{\mathcal{Q}}(y) = \frac{\lambda_{\mathcal{P}}\left[f^{-1}(y)\right]}{\left|f'\left[f^{-1}(y)\right]\right|} \tag{8.9}$$

($y \in \mathcal{Y}$).

Now, set the domain $\mathcal{X}$ to be the positive half-line, set $\mathcal{P} = \mathcal{E}_+$, and set $f(x) = \ln(x)$ ($x > 0$). For this setting $\lambda_{\mathcal{P}}(x) = \lambda_+(x) = c\epsilon x^{\epsilon-1}$, the domain $\mathcal{Y}$ is the real line, and Eq. (8.9) yields

$$\lambda_{\mathcal{Q}}(y) = \lambda_{\mathcal{P}}\left[\exp(y)\right]\exp(y) = c\epsilon \exp(\epsilon y) \tag{8.10}$$

($-\infty < y < \infty$). Analogously, set the domain $\mathcal{X}$ to be the positive half-line, set $\mathcal{P} = \mathcal{E}_-$, and set $f(x) = -\ln(x)$ ($x > 0$). For this setting $\lambda_{\mathcal{P}}(x) = \lambda_-(x) = c\epsilon x^{-\epsilon-1}$, the domain $\mathcal{Y}$ is the real line, and Eq. (8.9) yields

$$\lambda_{\mathcal{Q}}(y) = \lambda_{\mathcal{P}}\left[\exp(-y)\right]\exp(-y) = c\epsilon \exp(\epsilon y) \tag{8.11}$$

($-\infty < y < \infty$).

Proof of independence

Consider the Poisson process $\tilde{\mathcal{E}}$, with its exponential intensity $\tilde{\lambda}(y) = c\epsilon \exp(\epsilon y)$ ($-\infty < y < \infty$). In what follows, we set our focus on the first $m+1$ order statistics of this Poisson process: $\tilde{P}(1) < \tilde{P}(2) < \cdots < \tilde{P}(m+1)$.

The Poisson-process statistics imply that the probability density function of the random vector

$$\mathbf{Y} = \left(\tilde{P}(1), \tilde{P}(2), \ldots, \tilde{P}(m+1)\right) \tag{8.12}$$

is given by

$$f_{\mathbf{Y}}(y_1, y_2, \ldots, y_{m+1})$$

$$= \prod_{k=1}^{m+1}\left\{\exp\left[-\int_{y_{k-1}}^{y_k}\tilde{\lambda}(y')dy'\right]\cdot \tilde{\lambda}(y_k)\right\} \tag{8.13}$$

$$= \exp\left[-\int_{-\infty}^{y_{m+1}}\tilde{\lambda}(y')dy'\right]\cdot \prod_{k=1}^{m+1}\tilde{\lambda}(y_k) ,$$

where $-\infty = y_0 < y_1 < \cdots < y_{m+1}$.

Consider now the m consecutive gaps between the aforementioned order statistics: $\tilde{P}(2) - \tilde{P}(1), \tilde{P}(3) - \tilde{P}(2), \ldots, \tilde{P}(m+1) - \tilde{P}(m)$. The probability density function of the random vector

$$\mathbf{U} = \left(\tilde{P}(2) - \tilde{P}(1), \tilde{P}(3) - \tilde{P}(2), \ldots, \tilde{P}(m+1) - \tilde{P}(m)\right) \tag{8.14}$$

is given by

8 Exponent Estimation

$$f_U(u_1, u_2, \ldots, u_m)$$
$$= \int_{-\infty}^{\infty} f_Y(y, y+u_1, \ldots, y+u_1+\cdots+u_m) \, dy, \quad (8.15)$$

where $u_k > 0$ ($k = 1, \ldots, m$). Substituting Eq. (8.13) into Eq. (8.15), and using the shorthand notation $v_k = u_1 + \cdots + u_k$ ($k = 1, \ldots, m$), we have

$$f_U(u_1, u_2, \ldots, u_m)$$
$$= \int_{-\infty}^{\infty} \left\{ \exp\left[-\int_{-\infty}^{y+v_m} \tilde{\lambda}(y') \, dy'\right] \cdot \tilde{\lambda}(y) \cdot \prod_{k=1}^{m} \tilde{\lambda}(y+v_k) \right\} dy. \quad (8.16)$$

Set $\tilde{\Lambda}(y) = c \exp(\epsilon y)$ ($-\infty < y < \infty$), and note that

$$\exp\left[-\int_{-\infty}^{y+v_m} \tilde{\lambda}(y') \, dy'\right] = \exp\left[-\tilde{\Lambda}(y) \cdot \exp(\epsilon v_m)\right]. \quad (8.17)$$

Also note that
$$\tilde{\lambda}(y+v_k) = \tilde{\lambda}(y) \cdot \exp(\epsilon v_k) \quad (8.18)$$

($k = 1, \ldots, m$). Substituting Eqs. (8.17) and (8.18) into Eq. (8.16) yields:

$$f_U(u_1, u_2, \ldots, u_m)$$
$$= \int_{-\infty}^{\infty} \left\{ \exp\left[-\tilde{\Lambda}(y) \cdot \exp(\epsilon v_m)\right] \cdot \tilde{\lambda}(y)^{m+1} \cdot \prod_{k=1}^{m} \exp(\epsilon v_k) \right\} dy \quad (8.19)$$
$$= \prod_{k=1}^{m} \exp(\epsilon v_k) \cdot \int_{-\infty}^{\infty} \exp\left[-\tilde{\Lambda}(y) \cdot \exp(\epsilon v_m)\right] \tilde{\lambda}(y)^{m+1} \, dy.$$

Using the change of variables $x = \tilde{\Lambda}(y)$, noting that $\tilde{\lambda}(y) = \epsilon x$ and $dx = \tilde{\lambda}(y) \, dy$, and using the properties of the Gamma function, we have

$$\int_{-\infty}^{\infty} \exp\left[-\tilde{\Lambda}(y) \cdot \exp(\epsilon v_m)\right] \tilde{\lambda}(y)^{m+1} \, dy$$
$$= \int_0^{\infty} \exp\left[-x \cdot \exp(\epsilon v_m)\right] \cdot (\epsilon x)^m \, dx \quad (8.20)$$
$$= m! \epsilon^m \cdot \exp\left[-(m+1)\epsilon v_m\right].$$

Substituting Eq. (8.20) into Eq. (8.19) yields:

$$f_\mathbf{U}(u_1, u_2, \ldots, u_m)$$

$$= \prod_{k=1}^{m} \exp(\epsilon v_k) \cdot m! \epsilon^m \cdot \exp[-(m+1)\epsilon v_m]$$

$$= m! \epsilon^m \frac{\prod_{k=1}^{m} \exp[\epsilon(u_1 + \cdots + u_k)]}{\exp[(m+1)\epsilon(u_1 + \cdots + u_m)]} \qquad (8.21)$$

$$= m! \epsilon^m \frac{\prod_{k=1}^{m} \exp[\epsilon(m-k+1)u_k]}{\prod_{k=1}^{m} \exp[\epsilon(m+1)u_k]}$$

$$= m! \epsilon^m \prod_{k=1}^{m} \exp(-k\epsilon u_k)$$

$$= \prod_{k=1}^{m} k\epsilon \exp(-k\epsilon u_k) \ .$$

Equation (8.21) is the probability density function of the random vector $\mathbf{U}$ of Eq. (8.14). The product-form of this probability density function implies that: (i) the components of the random vector $\mathbf{U}$ are independent random variables; and (ii) the kth component of the random vector $\mathbf{U}$ is an exponential random variable with mean $1/(k\epsilon)$ ($k = 1, \ldots, m$).

References

1. J.F.C. Kingman, *Poisson Processes* (Oxford University Press, Oxford, 1993)
2. G.E. Moore, Electronics **38**, 114 (1965)
3. V. Vinge, in *Proceedings of a Symposium Vision-21: Interdisciplinary Science and Engineering in the Era of CyberSpace* (NASA Conference Publication CP-10129, 1993)
4. H. Moravec, J. Evol. Technol. **1**, 10 (1998)
5. R. Kurzweil, Law of accelerating returns (2001), http://www.kurzweilai.net/the-law-of-accelerating-returns
6. I. Eliazar, M.F. Shlesinger, Phys. A **494**, 430 (2018)
7. K. Krishnamoorthy, *Handbook of Statistical Distributions with Applications* (CRC Press, Boca Raton, 2016)
8. N.T. Thomopoulos, *Statistical Distributions* (Springer, New York, 2017)

Chapter 9
Socioeconomic Analysis

In this chapter, we observe the power Poisson processes, $\mathcal{E}_+$ and $\mathcal{E}_-$, from a socioeconomic perspective, and explore their socioeconomic behavior. As we shall see, this socioeconomic analysis will give rise to power Lorenz curves.

9.1 Socioeconomic Perspective

Our goal in this chapter is to analyze the power Poisson processes, $\mathcal{E}_+$ and $\mathcal{E}_-$, from a socioeconomic perspective. To that end, we consider the power Poisson processes $\mathcal{E}_+$ and $\mathcal{E}_-$ to represent human societies—with their points representing, respectively, the wealth values of the members of these societies. Then, in the context of this socioeconomic representation, we address the following question: how *equal/unequal* is the distribution of wealth in these human societies?

To assess the socioeconomic equality/inequality of a given human society we need to apply random sampling. Namely, we need to be able to pick at random a member of the society. As the power Poisson processes, $\mathcal{E}_+$ and $\mathcal{E}_-$, comprise of infinitely many points, we cannot apply random sampling directly. To circumvent this sampling obstacle, we resort to the threshold analysis of Chap. 4. Specifically, in the case of the power Poisson process $\mathcal{E}_+$, we consider the points that reside below a positive threshold level l. And, in the case of the power Poisson process $\mathcal{E}_-$, we consider the points that reside above a positive threshold level l.

For the power Poisson process $\mathcal{E}_+$ recall from Chap. 4 that there are finitely many points that reside below the threshold level l, and that these points are IID copies of the inverse-Pareto random variable $B_+(l)$. Moreover, the inverse-Pareto random variable $B_+(l)$ satisfies the following linear scaling:

$$B_+(l) = l \cdot B_+(1) , \qquad (9.1)$$

the equality being in law.

Similarly, for the power Poisson process $\mathcal{E}_-$ recall from Chap. 4 that there are finitely many points that reside above the threshold level l, and that these points are IID copies of the Pareto random variable $A_-(l)$. Moreover, the Pareto random variable $A_-(l)$ satisfies the following linear scaling:

$$A_-(l) = l \cdot A_-(1), \tag{9.2}$$

the equality being in law.

In what follows, we shall use the aforementioned random sampling, as well as the linear-scaling formulae of Eqs. (9.1) and (9.2).

9.2 Disparity Curve

Consider a human society comprising individuals with positive wealth values, and set W to be the wealth of a randomly sampled society member. Hence, W is a positive-valued random variable. Given two IID copies of the random variable W—denote them as W_1 and W_2—we introduce the following random variable:

$$\widehat{W} = \frac{W_1 - W_2}{W_1 + W_2}. \tag{9.3}$$

Namely, W_1 and W_2 are the wealth values of two society members that are sampled independently and at random. The random variable $\widehat{W}$ quantifies the wealth-disparity between the two randomly-sampled society members: their wealth-difference $W_1 - W_2$, measured relative to their combined wealth $W_1 + W_2$.

The random variable $\widehat{W}$ has the following features. (I) It takes values in the bounded range $-1 < \widehat{W} < 1$. (II) It is symmetric: $\widehat{W} = -\widehat{W}$, the equality being in law. (III) It has finite moments of all orders; in particular, it has a zero mean $\mathbf{E}[\widehat{W}] = 0$. (IV) It vanishes if and only if the random variable W is constant: $\widehat{W} = 0 \Leftrightarrow W = const$, where the equalities hold with probability one; in other words, the random variable $\widehat{W}$ vanishes if and only if the human society is communist—with all its members sharing a common wealth value. (V) It is invariant with respect to changes-of-scale: $\widehat{sW} = \widehat{W}$ for any positive scale parameter s; in other words, the random variable $\widehat{W}$ is invariant with respect to the specific currency via which wealth is measured. (VI) It is invariant with respect to reciprocation: $\widehat{1/W} = \widehat{W}$. These six features follow straightforwardly from Eq. (9.3).

Set $f(w)$ ($w > 0$) to denote the probability density function of the wealth random variable W. In turn, the probability density function of the wealth-disparity random variable $\widehat{W}$ is given by

9.2 Disparity Curve

$$\widehat{f}(u) = \begin{cases} \frac{2}{(1+u)^2} \int_0^\infty f\left(\frac{1-u}{1+u}w\right) f(w) \, w \, dw \\ \frac{2}{(1-u)^2} \int_0^\infty f\left(\frac{1+u}{1-u}w\right) f(w) \, w \, dw \end{cases} \quad (9.4)$$

($-1 < u < 1$); the derivation of Eq. (9.4) is detailed in the Methods. The symmetry feature (II) of the wealth-disparity random variable $\widehat{W}$ implies the symmetry of its density function: $\widehat{f}(-u) = \widehat{f}(u)$ ($-1 < u < 1$); the two lines on the right-hand side of Eq. (9.4) manifest this symmetry.

The greater the mass that the density function $\widehat{f}(u)$ concentrates near the origin ($u = 0$), the more *equal* the distribution of wealth. The greater the mass that the density function $\widehat{f}(u)$ concentrates near the endpoints ($u = \pm 1$), the more *unequal* the distribution of wealth. Hence, the shape of the density function $\widehat{f}(u)$ reflects the *socioeconomic disparity* of the human society under consideration. In what follows, we term the density function of Eq. (9.4) the "*disparity curve*" of the human society under consideration. The disparity curve, and its variants, have a broad range of scientific applications [1–12].

9.3 Disparity-Curve Analysis

With the disparity curve of Eq. (9.4) at hand, we are now in a position to investigate the inherent *socioeconomic disparity* of the power Poisson processes, $\mathcal{E}_+$ and $\mathcal{E}_-$. This section is based on [13], which explored the structure and the structural phase transitions of Poisson processes with power intensities.

As described in Sect. 9.1, we consider the human society to be the power Poisson process, $\mathcal{E}_+$, and focus on its points that reside below the positive threshold level l. In this case, the wealth random variable is $W = B_+(l)$. Combining together the linear-scaling property of Eq. (9.1) and the scale-invariance feature (V) of the wealth-disparity random variable $\widehat{W}$, we can set the wealth random variable to be $W = B_+(1)$. Hence, the corresponding disparity curve is *invariant* with respect to the threshold level l. Consequently, taking the limit $l \to \infty$ we conclude that: the disparity curve of the power Poisson process $\mathcal{E}_+$ is the disparity curve of the inverse-Pareto random variable $B_+(1)$.

Similarly, as described in Sect. 9.1, we consider the human society to be the power Poisson process $\mathcal{E}_-$, and focus on its points that reside above the positive threshold level l. In this case, the wealth random variable is $W = A_-(l)$. Combining together the linear-scaling property of Eq. (9.2) and the scale-invariance feature (V) of the wealth-disparity random variable $\widehat{W}$, we can set the wealth random variable to be $W = A_-(1)$. Hence, the corresponding disparity curve is *invariant* with respect to the threshold level l. Consequently, taking the limit $l \to 0$ we conclude that: the disparity curve of the power Poisson process $\mathcal{E}_-$ is the disparity curve of the Pareto random variable $A_-(1)$.

In Chap. 4 we noted the reciprocal connection between the inverse-Pareto random variable $B_+(1)$ and the Pareto random variable $A_-(1)$. Specifically: $B_+(1) = 1/A_-(1)$ and $A_-(1) = 1/B_+(1)$, the equalities being in law. Hence, the reciprocation-invariance feature (VI) of the wealth-disparity random variable $\widehat{W}$ implies that disparity curves of these two Pareto random variables coincide. In turn, the disparity curves of the power Poisson processes, $\mathcal{E}_+$ and $\mathcal{E}_-$, coincide.

The probability density function of the inverse-Pareto random variable $B_+(1)$ is $f(w) = \epsilon w^{\epsilon-1}$ ($0 < w < 1$), and the probability density function of the Pareto random variable $A_-(1)$ is $f(w) = \epsilon w^{-\epsilon-1}$ ($1 < w < \infty$). A calculation using Eq. (9.4) yields the following disparity curve of the power Poisson processes $\mathcal{E}_+$ and $\mathcal{E}_-$:

$$\widehat{f}(u) = \begin{cases} \epsilon(1+u)^{\epsilon-1}/(1-u)^{\epsilon+1} & -1 < u \leq 0, \\ \epsilon(1-u)^{\epsilon-1}/(1+u)^{\epsilon+1} & 0 \leq u < 1; \end{cases} \qquad (9.5)$$

the derivation of Eq. (9.5) is detailed in the Methods.

The disparity curve of Eq. (9.5) displays two different shapes, depending on the value of the exponent ϵ. (I) For exponent values $\epsilon < 1$, the disparity curve is unbounded and is W-shaped: it explodes at the endpoints ($u \to \pm 1$); it attains local minima at $u = \pm \epsilon$; and it attains a local maximum at the origin ($u = 0$). (II) For exponent values $\epsilon \geq 1$, the disparity curve is unimodal: it increases monotonically below the origin ($-1 < u < 0$); it peaks at the origin with $\widehat{f}(0) = \epsilon$; and it decreases monotonically above the origin ($0 < u < 1$). Thus, borrowing Mandelbrot's jargon [14–16], we conclude that: the power Poisson processes, $\mathcal{E}_+$ and $\mathcal{E}_-$, display a *wild socioeconomic disparity* in the exponent range $\epsilon < 1$, and a *mild socioeconomic disparity* in the exponent range $\epsilon \geq 1$. The exponent value $\epsilon = 1$ is a tipping point at which a *disparity phase transition* takes place.

9.4 Lorenz Curves

As noted in the introduction, the method of *Lorenz curves* is widely applied in economics and in the social sciences to quantify the distribution of income and wealth in human societies [17–20]. We present now a brief description of this method, which is based on [21, 22].

Similarly to Sect. 9.2, consider a human society comprising individuals with non-negative wealth values. The Lorenz curve $y = L(x)$ ($0 \leq x, y \leq 1$) of the human society is defined as follows: the low (poor) $100x\%$ of the society's population are in possession of $100y\%$ of the society's overall wealth. Alternatively, the Lorenz curve $y = \bar{L}(x)$ ($0 \leq x, y \leq 1$) of the human society is defined as follows: the high (rich) $100x\%$ of the society's population are in possession of $100y\%$ of the society's overall wealth. By their very definitions, the Lorenz curves are coupled: $L(x) + \bar{L}(1-x) = 1$ ($0 \leq x \leq 1$); hence, each of the two Lorenz curves fully determines the other.

9.4 Lorenz Curves

The two Lorenz curves reside in the unit square ($0 \leq x, y \leq 1$), and are non-decreasing functions that "start" from the zero level $L(0) = 0 = \bar{L}(0)$, and that "end" at the unit level $L(1) = 1 = \bar{L}(1)$. The Lorenz curve $y = L(x)$ is a convex function, and the Lorenz curve $y = \bar{L}(x)$ is a concave function. Consequently, the Lorenz curves are ordered as follows:

$$0 \leq L(x) \leq x \leq \bar{L}(x) \leq 1 \tag{9.6}$$

($0 \leq x \leq 1$).

Namely, the Lorenz curve $y = L(x)$ is bounded from below by the "floor" $y = 0$ of the unit square, and is bounded from above by the diagonal line $y = x$ of the unit square. And, the Lorenz curve $y = \bar{L}(x)$ is bounded from below by the diagonal line $y = x$ of the unit square, and is bounded from above by the "ceiling" $y = 1$ of the unit square. These bounds manifest the two socioeconomic extremes of human societies: perfect equality and perfect inequality.

Perfect equality is the case of a purely egalitarian distribution of wealth, i.e., a communist society. In terms of the Lorenz curves, perfect equality is characterized by the diagonal line of the unit square: $L(x) = x = \bar{L}(x)$ ($0 \leq x \leq 1$). In the context of societies with infinitely large populations, perfect inequality is the case where an "oligarchy" that constitutes 0% of the society's population is in possession of 100% of the society's overall wealth. In terms of the Lorenz curves perfect inequality is characterized by the "floor" and the "ceiling" of the unit square: $L(x) = 0$ ($0 \leq x < 1$) and $\bar{L}(x) = 1$ ($0 < x \leq 1$).

The closer the Lorenz curves are to the diagonal line of the unit square—the more equal the distribution of wealth. Conversely, the further away the Lorenz curves are from the diagonal line of the unit square—the more unequal the distribution of wealth. Thus, the *geometric distance* of the Lorenz curves from the diagonal line quantifies the *socioeconomic inequality* of the human society under consideration.

As in Sect. 9.2, we sample at random a member of the human society under consideration, and set W to be the wealth of this randomly-sampled member. Also, we denote by $F(w) = \Pr(W < w)$ ($w > 0$) the corresponding cumulative distribution function, and by $\bar{F}(w) = \Pr(W > w)$ ($w > 0$) the corresponding tail distribution function. Namely, for a given wealth value w the following holds: $F(w)$ is the proportion of the society members with wealth smaller than w, and $\bar{F}(w)$ is the proportion of the society members with wealth larger than w.

In terms of the inverse of the cumulative distribution function, $F^{-1}(u)$ ($0 < u < 1$), the Lorenz curve $y = L(x)$ admits the following *Pietra representation* [23]:

$$L(x) = \frac{1}{\mathbf{E}[W]} \int_0^x F^{-1}(u) \, du \tag{9.7}$$

($0 \leq x \leq 1$). And, in terms of the inverse of the tail distribution function, $\bar{F}^{-1}(u)$ ($0 < u < 1$), the Lorenz curve $y = \bar{L}(x)$ admits the following *Pietra representation* [23]:

$$\bar{L}(x) = \frac{1}{\mathbf{E}[W]} \int_0^x \bar{F}^{-1}(u)\,du \qquad (9.8)$$

($0 \leq x \leq 1$). For the derivations of the Pietra representations of Eqs. (9.7) and (9.8) consult, for example, [24].

Equations (9.7) and (9.8) facilitate the calculation of the Lorenz curves of *any* non-negative random variable W with a positive mean, $0 < \mathbf{E}[W] < \infty$. For a non-negative random variable W with a divergent mean, $\mathbf{E}[W] = \infty$, the perfect-inequality Lorenz curves hold: $L(x) = 0$ ($0 \leq x < 1$) and $\bar{L}(x) = 1$ ($0 < x \leq 1$). Also, Eqs. (9.7) and (9.8) imply invariance with respect to changes-of-scale: replacing the random variable W by the random variable sW, where s is a positive scale parameter, does not affect the Lorenz curves. Indeed, by their very definitions, the Lorenz curves are invariant with respect to the specific currency via which wealth is measured.

9.5 Lorenz-Curves Analysis

With the method of Lorenz curves at hand, we are now in position to investigate the inherent *socioeconomic inequality* of the power Poisson processes $\mathcal{E}_+$ and $\mathcal{E}_-$. As Sect. 9.3, this section is also based on [13].

As described in Sect. 9.1, we consider the human society to be the power Poisson process $\mathcal{E}_+$, and focus on its points that reside below the positive threshold level l. In this case, the wealth random variable is $W = B_+(l)$. Combining together the linear-scaling property of Eq. (9.1), and the scale-invariance feature of the Lorenz curves, we can set the wealth random variable to be $W = B_+(1)$. Hence, the corresponding Lorenz curves are *invariant* with respect to the threshold level l. Consequently, taking the limit $l \to \infty$ we conclude that: the Lorenz curves of the power Poisson process $\mathcal{E}_+$ are the Lorenz curves of the inverse-Pareto random variable $B_+(1)$.

Similarly, as described in Sect. 9.1, we consider the human society to be the power Poisson process $\mathcal{E}_-$, and focus on its points that reside above the positive threshold level l. In this case, the wealth random variable is $W = A_-(l)$. Combining together the linear-scaling property of Eq. (9.2), and the scale-invariance feature of the Lorenz curves, we can set the wealth random variable to be $W = A_-(1)$. Hence, the corresponding Lorenz curves are *invariant* with respect to the threshold level l. Consequently, taking the limit $l \to 0$ we conclude that: the Lorenz curves of the power Poisson process $\mathcal{E}_-$ are the Lorenz curves of the Pareto random variable $A_-(1)$.

The cumulative distribution function of the inverse-Pareto random variable $B_+(1)$ is $F(w) = w^\epsilon$ ($0 < w < 1$). A straightforward calculation using Eq. (9.7) yields the following Lorenz curve of the power Poisson process $\mathcal{E}_+$:

$$L_+(x) = x^{1+1/\epsilon} \qquad (9.9)$$

($0 \leq x \leq 1$). Note that the term $1/\epsilon$ appearing on the right-hand side of Eq. (9.9) is the *slope* of the *inverse-Zipf law* that corresponds to the order statistics of the power Poisson process $\mathcal{E}_+$ (see Eq. (7.9) in Chap. 7).

The tail distribution function of the Pareto random variable $A_-(1)$ is $\bar{F}(w) = 1/w^\epsilon$ ($1 < w < \infty$). Consequently, a straightforward calculation using Eq. (9.8) yields the following Lorenz curve of the power Poisson process $\mathcal{E}_-$:

$$\bar{L}_-(x) = x^{1-1/\epsilon} \tag{9.10}$$

($0 \leq x \leq 1$) for exponents $\epsilon > 1$; and $\bar{L}_-(x) = 1$ ($0 < x \leq 1$) for exponents $\epsilon \leq 1$. Note that the term $-1/\epsilon$ appearing on the right-hand side of Eq. (9.10) is the *slope* of the *Zipf law* that corresponds to the order statistics of the power Poisson process $\mathcal{E}_-$ (see Eq. (7.11) in Chap. 7).

The Lorenz curve of Eq. (9.9) characterizes a *power Lorenz curve* $y = L(x)$; note that the exponent $1 + \frac{1}{\epsilon}$ of this power Lorenz curve is greater than one—in accord with the convexity property of the Lorenz curve $y = L(x)$. The Lorenz curve of Eq. (9.10) characterizes a *power Lorenz curve* $y = \bar{L}(x)$; note that the exponent $1 - \frac{1}{\epsilon}$ of this power Lorenz curve is smaller than one—in accord with the concavity property of the Lorenz curve $y = \bar{L}(x)$.

To highlight the effect of the exponent ϵ on the Lorenz curves of the power Poisson processes $\mathcal{E}_+$ and $\mathcal{E}_-$ we introduce, respectively, the notations $L_+(x; \epsilon)$ and $\bar{L}_-(x; \epsilon)$. With these notations, we note three key properties of the Lorenz curves. (I) The power Poisson processes, $\mathcal{E}_+$ and $\mathcal{E}_-$, reach the perfect-inequality extreme in the limits $\epsilon \to 0$ and $\epsilon \to 1$, respectively: $\lim_{\epsilon \to 0} L_+(x; \epsilon) = 0$ ($0 \leq x < 1$) and $\lim_{\epsilon \to 1} \bar{L}_-(x; \epsilon) = 1$ ($0 < x \leq 1$). (II) The larger the exponent ϵ, the more equal the power Poisson processes, $\mathcal{E}_+$ and $\mathcal{E}_-$, i.e., if $\epsilon_2 > \epsilon_1$ then: $L_+(x; \epsilon_2) > L_+(x; \epsilon_1)$ ($0 < x < 1; \epsilon_1 > 0$) and $\bar{L}_-(x; \epsilon_2) < \bar{L}_-(x; \epsilon_1)$ ($0 < x < 1; \epsilon_1 > 1$). (III) The power Poisson processes, $\mathcal{E}_+$ and $\mathcal{E}_-$, reach the perfect-equality extreme in the limit $\epsilon \to \infty$, i.e.: $\lim_{\epsilon \to \infty} L_+(x; \epsilon) = x = \lim_{\epsilon \to \infty} \bar{L}_-(x; \epsilon)$ ($0 \leq x \leq 1$).

9.6 Inequality Indices

The disparity curve described in Sect. 9.2, and the Lorenz curves described in Sect. 9.4, are both functions. As such, these curves provide infinite-dimensional quantifications of the distribution of wealth in a human society under consideration—and this may often be regarded as "to much information". Alternative to these curves are *inequality indices*—which are widely applied in economics and in the social sciences [25–28], and which are applicable well beyond these fields [29, 30].

An inequality index $\mathcal{I}$ gives a score to the socioeconomic inequality of a human society under consideration. The score takes values in the unit interval, $0 \leq \mathcal{I} \leq 1$. The zero score, $\mathcal{I} = 0$, characterizes the socioeconomic extreme of perfect equality.

The unit score, $\mathcal{I} = 1$, is attained in the socioeconomic extreme of perfect inequality.[1] Also, as the disparity curve and the Lorenz curves, the inequality index $\mathcal{I}$ is scale-invariant: a change-of-scale of the underlying wealth random variable W does not affect the index. In other words, the score $\mathcal{I}$ is invariant with respect to the specific currency via which wealth is measured.

Consider the power Poisson processes, $\mathcal{E}_+$ and $\mathcal{E}_-$, and a general inequality index $\mathcal{I}$. On the one hand, as noted in Sect. 9.1, the Pareto random variables $B_+(l)$ and $A_-(l)$—emanating from the power Poisson processes $\mathcal{E}_+$ and $\mathcal{E}_-$ via the introduction of the threshold level l—display the linear-scaling properties of Eqs. (9.1) and (9.2), respectively. On the other hand, as noted above, the inequality index $\mathcal{I}$ is scale-invariant. Hence, as argued for the disparity curve of Sect. 9.2, as well as for the Lorenz curves of Sect. 9.4, the following conclusions hold. (I) The inequality index $\mathcal{I}$ of the power Poisson process $\mathcal{E}_+$ is the inequality index $\mathcal{I}$ of the inverse-Pareto random variable $B_+(1)$. (II) The inequality index $\mathcal{I}$ of the power Poisson process $\mathcal{E}_-$ is the inequality index $\mathcal{I}$ of the Pareto random variable $A_-(1)$.

In the next two sections, we shall focus on two specific inequality indices: the *Gini index* and the *reciprocation index*. These indices will demonstrate how the inherent socioeconomic inequality of the power Poisson processes, $\mathcal{E}_+$ and $\mathcal{E}_-$, depends on their exponent ϵ.

9.7 Gini Index

Perhaps the most popular inequality index is the *Gini index* [31–36], which we denote $\mathcal{G}$. In this section, we describe the relation of the Gini index to the disparity curve and to the Lorenz curves, and then present the Gini indices of the power Poisson processes, $\mathcal{E}_+$ and $\mathcal{E}_-$. In what follows, we use the notations and settings of Sects. 9.2 and 9.4.

The Gini index is related to the disparity curve as follows. The disparity curve is based on the wealth-disparity random variable of Eq. (9.3): $\widehat{W} = (W_1 - W_2) / (W_1 + W_2)$, where the random variables W_1 and W_2 are the wealth values of two society members that are sampled independently and at random. The random variable $\widehat{W}$ is the symmetrization of the random variable $|W_1 - W_2| / (W_1 + W_2)$, which manifests the following ratio: the absolute-value wealth-difference $|W_1 - W_2|$, measured relative to the combined wealth $W_1 + W_2$.

Now, consider the ratio $|W_1 - W_2| / (W_1 + W_2)$. The mean of the nominator, $\mathbf{E}[|W_1 - W_2|]$, is the mean absolute difference (MAD) between the random variables W_1 and W_2 [37–39]. The mean of the denominator, $\mathbf{E}[W_1 + W_2]$, is twice the mean of the wealth random variable, $2\mathbf{E}[W]$. In terms of these two means, the Gini index admits the following ratio formulation [30]:

---

[1] In the case of socio-geometric inequality indices, the unit score $\mathcal{I} = 1$ characterizes the socioeconomic extreme of perfect inequality [21, 22]. However, in the case of general inequality indices the unit score may be attained also in scenarios other than perfect inequality.

9.7 Gini Index

$$\mathcal{G} = \frac{\mathbf{E}[|W_1 - W_2|]}{2\mathbf{E}[W]} . \tag{9.11}$$

Equation (9.11) implies that the Gini index $\mathcal{G}$ is a "mean version" of the wealth-disparity random variable $\widehat{W}$ of Eq. (9.3). Equation (9.11) also implies the scale invariance of the Gini index $\mathcal{G}$: replacing the wealth random variable W by the random variable sW, where s is a positive scale parameter, does not affect the index.

The Gini index is related to the Lorenz curves as follows. In Sect. 9.2, we noted that the further away the Lorenz curves $y = L(x)$ and $y = \bar{L}(x)$ are from the diagonal line $y = x$ of the unit square, the more unequal the distribution of wealth. Hence, one can use the areas captured between the Lorenz curves and the diagonal line as a geometric measure of socioeconomic inequality. In terms of these two areas, the Gini index admits the following formulations [30]:

$$\mathcal{G} = 2\int_0^1 [x - L(x)]\,dx = 2\int_0^1 \left[\bar{L}(x) - x\right]dx . \tag{9.12}$$

It is evident from Eq. (9.12) that the Gini index yields the zero score if and only if the Lorenz curves coincide with the diagonal line $y = x$ of the unit square: $\mathcal{G} = 0 \Leftrightarrow L(x) = x = \bar{L}(x)$ ($0 \leq x \leq 1$). Also, it is evident from Eq. (9.12) that the Gini index yields the unit score if and only if the Lorenz curves coincide with the "floor" and the "ceiling" of the unit square: $\mathcal{G} = 1 \Leftrightarrow L(x) = 0$ and $\bar{L}(x) = 1$ ($0 < x < 1$).

As established in Sect. 9.4, the Lorenz curve of the power Poisson process $\mathcal{E}_+$ is given by: $L_+(x) = x^{1+1/\epsilon}$ ($0 \leq x \leq 1$). Substituting this Lorenz curve into Eq. (9.12), a straightforward calculation yields the Gini index

$$\mathcal{G}_+ = \frac{1}{2\epsilon + 1} . \tag{9.13}$$

As established in Sect. 9.4, the Lorenz curve of the power Poisson process $\mathcal{E}_-$ is given by $\bar{L}_-(x) = x^{1-1/\epsilon}$ ($0 \leq x \leq 1$) for exponents $\epsilon > 1$; and $\bar{L}_-(x) = 1$ ($0 < x \leq 1$) for exponents $\epsilon \leq 1$. Substituting this Lorenz curve into Eq. (9.12), a straightforward calculation yields the Gini index

$$\mathcal{G}_- = \frac{1}{2\epsilon - 1} \tag{9.14}$$

for exponents $\epsilon > 1$; and $\mathcal{G}_- = 1$ for exponents $\epsilon \leq 1$.

To highlight the effect of the exponent ϵ on the Gini indices of the power Poisson processes $\mathcal{E}_+$ and $\mathcal{E}_-$ we introduce, respectively, the notations $\mathcal{G}_+(\epsilon)$ and $\mathcal{G}_-(\epsilon)$. With these notations we observe that the Gini indices echo the three key properties of the corresponding Lorenz curves. (I) The power Poisson processes, $\mathcal{E}_+$ and $\mathcal{E}_-$, reach the perfect-inequality extreme in the limits $\epsilon \to 0$ and $\epsilon \to 1$, respectively: $\lim_{\epsilon \to 0} \mathcal{G}_+(\epsilon) = 1$ and $\lim_{\epsilon \to 1} \mathcal{G}_-(\epsilon) = 1$. (II) The larger the exponent ϵ, the more equal the power Poisson processes, $\mathcal{E}_+$ and $\mathcal{E}_-$, i.e., the Gini indices $\mathcal{G}_+(\epsilon)$ and $\mathcal{G}_-(\epsilon)$ are monotone decreasing with respect to the exponent ϵ. (III) The power Poisson processes, $\mathcal{E}_+$ and $\mathcal{E}_-$, reach the perfect-equality extreme in the limit $\epsilon \to \infty$, i.e.: $\lim_{\epsilon \to \infty} \mathcal{G}_+(\epsilon) = 0 = \lim_{\epsilon \to \infty} \mathcal{G}_-(\epsilon)$.

9.8 Reciprocation Index

An inequality index that is altogether different from the Gini index is the *reciprocation index* [30], which we denote by $\mathcal{R}$. In terms of the wealth random variable W, the reciprocation index is given by [30]:

$$\mathcal{R} = 1 - \frac{1}{\mathbf{E}[W]\mathbf{E}[1/W]}. \tag{9.15}$$

The reciprocation index belongs to a continuum of inequality indices which are based on the *Renyi entropy*, and which are termed "*Renyi spectrum*" [40, 41].

Jensen's inequality [42] assures that the reciprocation index takes values in the unit interval, $0 \leq \mathcal{R} \leq 1$. Jensen's inequality further assures that the reciprocation index yields the zero score only in the perfect-equality extreme: $\mathcal{R} = 0 \Leftrightarrow W = const$, where the equalities hold with probability one. The scale invariance of the reciprocation index $\mathcal{R}$ is evident from Eq. (9.15): replacing the random variable W by the random variable sW, where s is a positive scale parameter, does not affect the index. Also, it is evident from Eq. (9.15) that the reciprocation index $\mathcal{R}$ is invariant with respect to reciprocation: replacing the random variable W by the random variable $1/W$ does not affect the index.

Setting $W = B_+(1)$ in Eq. (9.15) yields the reciprocation index $\mathcal{R}$ of the power Poisson process $\mathcal{E}_+$, and setting $W = A_-(1)$ in Eq. (9.15) yields the reciprocation index $\mathcal{R}$ of the power Poisson process $\mathcal{E}_-$. In Chap. 4, we noted the reciprocal connection between the inverse-Pareto random variable $B_+(1)$ and the Pareto random variable $A_-(1)$. Specifically: $B_+(1) = 1/A_-(1)$ and $A_-(1) = 1/B_+(1)$, the equalities being in law. Hence, due to the reciprocation-invariance of the reciprocation index $\mathcal{R}$, the values of this index for the random variables $B_+(1)$ and $A_-(1)$ coincide. Consequently, the reciprocation indices of the power Poisson processes, $\mathcal{E}_+$ and $\mathcal{E}_-$, coincide.

In Chap. 4, we saw that the mean of the inverse-Pareto random variable is $\mathbf{E}[B_+(1)] = \frac{\epsilon}{\epsilon+1}$, and that the mean of the Pareto random variable is $\mathbf{E}[A_-(1)] = \frac{\epsilon}{\epsilon-1}$ (for exponents $\epsilon > 1$). Consequently, exploiting the reciprocal connection between the two Pareto random variables, a straightforward calculation using Eq. (9.15) yields the following reciprocation index of the power Poisson processes $\mathcal{E}_+$ and $\mathcal{E}_-$:

$$\mathcal{R} = \frac{1}{\epsilon^2} \tag{9.16}$$

for exponents $\epsilon > 1$; and $\mathcal{R} = 1$ for exponents $\epsilon \leq 1$. Note that the right-hand side of Eq. (9.16) is the square of the *slopes* of the *Zipf laws* that corresponds to the order statistics of the power Poisson processes $\mathcal{E}_+$ and $\mathcal{E}_-$ (see Eqs. (7.9) and (7.11) in Chap. 7).

To highlight the effect of the exponent ϵ on the reciprocation index of Eq. (9.16) we introduce the notation $\mathcal{R}(\epsilon)$, and observe the three following properties of the index. (I) It reaches the unit score in the limit $\epsilon \to 1$, i.e., $\lim_{\epsilon \to 1} \mathcal{R}(\epsilon) = 1$. (II) It is monotone decreasing with respect to the exponent ϵ. (III) It reaches the perfect-equality extreme in the limit $\epsilon \to \infty$, i.e., $\lim_{\epsilon \to \infty} \mathcal{R}(\epsilon) = 0$.

9.9 Summary

In this chapter, we explored the socioeconomic behavior of the power Poisson processes $\mathcal{E}_+$ and $\mathcal{E}_-$. To do so, we applied a socioeconomic perspective in which the power Poisson processes, $\mathcal{E}_+$ and $\mathcal{E}_-$, represent human societies; in turn, the points of these processes represent the wealth values of the members of these societies, respectively. With regard to this representation, we addressed the socioeconomic equality/inequality of the power Poisson processes $\mathcal{E}_+$ and $\mathcal{E}_-$.

Four socioeconomic quantifications of the power Poisson processes, $\mathcal{E}_+$ and $\mathcal{E}_-$, were analyzed. The first and second quantifications were the disparity curve and the Lorenz curves. The third and fourth quantifications were inequality indices: the popular Gini index, and the reciprocation index. These quantifications are well defined in the context of positive-valued random variables—yet not necessarily in the context of general Poisson processes over the positive half-line. Using the threshold analysis of Chap. 4, we established that these quantifications are well defined also in the context of the power Poisson processes $\mathcal{E}_+$ and $\mathcal{E}_-$.

The power Poisson processes, $\mathcal{E}_+$ and $\mathcal{E}_-$, share a common disparity curve and a common reciprocation index. For exponents $\epsilon < 1$, the common disparity curve exhibits wild socioeconomic disparity, and for exponents $\epsilon \geq 1$ it exhibits mild socioeconomic disparity. For exponents $\epsilon > 1$, the common reciprocation index is the square of the "Zipf slopes" of the power Poisson processes $\mathcal{E}_+$ and $\mathcal{E}_-$, and for exponents $\epsilon \leq 1$ it hits its unit-score upper bound.

The power Poisson processes, $\mathcal{E}_+$ and $\mathcal{E}_-$, have different Lorenz curves and different Gini indices. The Lorenz curves and the Gini index of the power Poisson process $\mathcal{E}_+$ exhibit the same qualitative behavior for all exponents $\epsilon > 0$. The Lorenz curves and the Gini index of the power Poisson process $\mathcal{E}_-$—for exponents $\epsilon \leq 1$—characterize the socioeconomic extreme of perfect inequality.

9.10 Methods

9.10.1 Equation (9.4)

Set $R = W_2/W_1$ to be the ratio of the IID random variables W_1 and W_2. Equation (9.3) implies that

$$\widehat{W} = \frac{W_1 - W_2}{W_1 + W_2} = \frac{1 - W_2/W_1}{1 + W_2/W_1} = \frac{1 - R}{1 + R}. \quad (9.17)$$

It follows from Eq. (9.17) that

$$\widehat{W} < u \Leftrightarrow R > \frac{1 - u}{1 + u} \quad (9.18)$$

$(-1 < u < 1)$, and hence:

$$\Pr\left(\widehat{W} < u\right) = \Pr\left(R > \frac{1 - u}{1 + u}\right) \quad (9.19)$$

$(-1 < u < 1)$. The ratio R is a positive-valued random variable, and we set $g(x)$ $(x > 0)$ to denote its probability density function. Differentiating Eq. (9.19) implies that the probability density function of the random variable $\widehat{W}$ is given by

$$\widehat{f}(u) = g\left(\frac{1-u}{1+u}\right) \frac{2}{(1+u)^2} \quad (9.20)$$

$(-1 < u < 1)$.

As the random variables W_1 and W_2 are IID with probability density function $f(w)$ $(w > 0)$, the cumulative distribution function of the ratio R is given by

$$\Pr(R < x) = \Pr\left(\tfrac{W_2}{W_1} < x\right)$$
$$= \Pr(W_2 < xW_1) \quad (9.21)$$
$$= \int_0^\infty \Pr(W_2 < xw) f(w)\, dw$$

$(x > 0)$. Differentiating Eq. (9.21) implies that the probability density function of the random variable R is given by

$$g(x) = \int_0^\infty f(xw)\, w f(w)\, dw \quad (9.22)$$

$(x > 0)$.

Combining Eqs. (9.20) and (9.22) yields

$$\widehat{f}(u) = \frac{2}{(1+u)^2} \int_0^\infty f\left(\frac{1-u}{1+u}w\right) wf(w)\, dw \qquad (9.23)$$

($-1 < u < 1$). The symmetry of the random variable $\widehat{W}$ (i.e. $\widehat{W} = -\widehat{W}$, the equality being in law) implies the symmetry of its probability density function: $\widehat{f}(-u) = \widehat{f}(u)$ ($-1 < u < 1$). Combined together with Eq. (9.23), this symmetry implies that

$$\widehat{f}(u) = \frac{2}{(1-u)^2} \int_0^\infty f\left(\frac{1+u}{1-u}w\right) wf(w)\, dw \qquad (9.24)$$

($-1 < u < 1$). Alternatively, Eq. (9.24) follows from Eq. (9.23) via the change-of-variables $w \mapsto w' = \frac{1+u}{1-u}w$. Equations (9.23) and (9.24) prove Eq. (9.4).

9.10.2 Equation (9.5)

Consider the probability density function $f_+(w) = \epsilon w^{\epsilon-1}$ ($0 < w < 1$). Substituting $f(w) = f_+(w)$ ($0 < w < 1$) and $f(w) = 0$ ($1 \leq w < \infty$) into Eq. (9.23), and considering $0 \leq u < 1$, we have

$$\begin{aligned}
\widehat{f}_+(u) &= \tfrac{2}{(1+u)^2} \int_0^1 f_+\left(\tfrac{1-u}{1+u}w\right) wf_+(w)\, dw \\
&= \tfrac{2}{(1+u)^2} \int_0^1 \epsilon \left(\tfrac{1-u}{1+u}w\right)^{\epsilon-1} w\epsilon w^{\epsilon-1} dw \\
&= \tfrac{2}{(1+u)^2} \left(\tfrac{1-u}{1+u}\right)^{\epsilon-1} \epsilon^2 \int_0^1 w^{2\epsilon-1} dw \\
&= \tfrac{2}{(1+u)^2} \left(\tfrac{1-u}{1+u}\right)^{\epsilon-1} \tfrac{\epsilon}{2} = \epsilon \tfrac{(1-u)^{\epsilon-1}}{(1+u)^{\epsilon+1}}
\end{aligned} \qquad (9.25)$$

($0 \leq u < 1$). In turn, using the symmetry of its probability density function $\widehat{f}(u)$, and considering $-1 < u \leq 0$, Eq. (9.25) implies that

$$\widehat{f}_+(u) = \epsilon \frac{(1+u)^{\epsilon-1}}{(1-u)^{\epsilon+1}} \qquad (9.26)$$

($-1 < u \leq 0$). Equations (9.25) and (9.26) yield Eq. (9.5).

References

1. I. Eliazar, Phys. A **356**, 207 (2005)
2. G. Oshanin, S. Redner, Europhys. Lett. **85**, 10008 (2009)
3. I. Eliazar, I.M. Sokolov, J. Phys. A **43**, 055001 (2010)

4. I.M. Sokolov, I. Eliazar, Phys. Rev. E **81**, 026107 (2010)
5. C. Mejia-Monasterio, G. Oshanin, G. Schehr, J. Stat. Mech. P06022 (2011)
6. G. Oshanin, Yu. Holovatch, G. Schehr, Phys. A **390**, 4340 (2011)
7. C. Mejia-Monasterio, G. Oshanin, G. Schehr, Phys. Rev. E **84**, 035203 (2011)
8. G. Oshanin, G. Schehr, Quant. Financ. **12**, 1325 (2012)
9. D. Boyer, D.S. Dean, C. Mejia-Monasterio, G. Oshanin, Phys. Rev. E **85**, 031136 (2012)
10. I. Eliazar, I.M. Sokolov, Phys. A **391**, 3043 (2012)
11. T.G. Mattos, C. Mejia-Monasterio, R. Metzler, G. Oshanin, Phys. Rev. E **86**, 031143 (2012)
12. G. Oshanin, A. Rosso, G. Schehr, Phys. Rev. Lett. **110**, 100602 (2013)
13. I. Eliazar, G. Oshanin, J. Phys. A: Math. Theor. **45**, 405003 (2012)
14. B.B. Mandelbrot, *Fractals and Scaling in Finance* (Springer, New York, 2013)
15. B. Mandelbrot, N.N. Taleb, in *The Known, the Unknown and the Unknowable in Financial Institutions*, ed. by F. Diebold, N. Doherty, R. Herring (Princeton University Press, Princeton, 2010), pp. 47–58
16. I. Eliazar, M.H. Cohen, Chaos Solitons Fractals **74**, 3 (2015)
17. M.O. Lorenz, Pub. Am. Stat. Assoc. **9**, 209 (1905)
18. J.L. Gastwirth, Econometrica **39**, 1037 (1971)
19. G.M. Giorgi, Metron **63**, 299 (2005)
20. D. Chotikapanich (ed.), *Modeling Income Distributions and Lorenz Curves* (Springer, New York, 2008)
21. I. Eliazar, Phys. A **426**, 93 (2015)
22. I. Eliazar, Phys. A **426**, 116 (2015)
23. G. Pietra, Atti del Reale Istituto Veneto di Scienze. Lettere ed Arti **LXXIV**, parte II, 775 (1914–1915)
24. I. Eliazar, I.M. Sokolov, Phys. A **391**, 1323 (2012)
25. P.B. Coulter, *Measuring Inequality: A Methodological Handbook* (Westview Press, Boulder, 1989)
26. G. Betti, A. Lemmi (eds.), *Advances on Income Inequality and Concentration Measures* (Routledge, New York, 2008)
27. L. Hao, D.Q. Naiman, *Assessing Inequality* (Sage, Los Angeles, 2010)
28. F. Cowell, *Measuring Inequality* (Oxford University Press, Oxford, 2011)
29. I. Eliazar, Phys. Rep. **649**, 1 (2016)
30. I. Eliazar, Ann. Phys. **389**, 306 (2018)
31. C. Gini, Sulla misura della concentrazione e della variabilita dei caratteri (1914) Atti del Reale Istituto Veneto di Scienze. Lettere ed Arti, a.a. 1913–1914 **LXXIII**, parte II, 1203–1248. English translation: Metron **63**, 3 (2005)
32. C. Gini, Econ. J. **31**, 124 (1921)
33. S. Yitzhaki, Res. Econ. Inequal. **8**, 13 (1998)
34. S. Yitzhaki, E. Schechtman, *The Gini Methodology* (Springer, New York, 2012)
35. G.M. Giorgi, S. Gubbiotti, Int. Stat. Rev. (2016). https://doi.org/10.1111/insr.12196
36. G.M. Giorgi, C. Gigliarano, J. Econ. Surv. (2016). https://doi.org/10.1111/joes.12185
37. C. Gini, Variabilita e mutabilita (1912), reprinted in: E. Pizetti, T. Salvemini, Memorie di metodologica statistica (Libreria Eredi Virgilio Veschi, Rome, 1955)
38. S. Yitzhaki, Metron **61**, 285 (2003)
39. S. Yitzhaki, P.J. Lambert, Metron **71**, 97 (2013)
40. I. Eliazar, Phys. A **481**, 90 (2017)
41. I. Eliazar, Phys. A **469**, 824 (2017)
42. J.L.W.V. Jensen, Acta Math. **30**, 175 (1906)

Chapter 10
Fractality

In this chapter, we explore the fractal properties of the power Poisson processes $\mathcal{E}_+$ and $\mathcal{E}_-$. Each fractal property will be based on invariance with respect to a certain perspective.

10.1 Scale Invariance

A key feature of fractals is that they essentially "look the same" under different scales [1–3]: zooming in and zooming out yields, in essence, the same intrinsic pattern. In this section, we consider the change-of-scale $x \mapsto s \cdot x$ ($x > 0$), where s is a positive scale parameter.

A result from the theory of Poisson processes (the "displacement theorem" [4]) implies that applying the change-of-scale to the points of the power Poisson process $\mathcal{E}_+$ yields a new Poisson process $\tilde{\mathcal{E}}_+$, also with positive-valued points. Moreover, in terms of the intensity function $\lambda_+(x) = c\epsilon x^{\epsilon-1}$ of the Poisson process $\mathcal{E}_+$, the intensity function of the new Poisson process is given by

$$\tilde{\lambda}_+(x) = s^{-\epsilon} \cdot \lambda_+(x) \tag{10.1}$$

($x > 0$); the derivation of Eq. (10.1) is detailed in the Methods. Equation (10.1) implies that the change-of-scale affects only the *magnitude* of the intensity function $\lambda_+(x)$, and does not affect its *shape*.

The aforementioned result from the theory of Poisson processes implies that applying the change-of-scale to the points of the power Poisson process $\mathcal{E}_-$ yields a new Poisson process $\tilde{\mathcal{E}}_-$, also with positive-valued points. Moreover, in terms of the intensity function $\lambda_-(x) = c\epsilon x^{-\epsilon-1}$ of the Poisson process $\mathcal{E}_-$, the intensity function of the new Poisson process is given by

$$\tilde{\lambda}_-(x) = s^\epsilon \cdot \lambda_-(x) \tag{10.2}$$

($x > 0$); the derivation of Eq. (10.2) is detailed in the Methods. Similar to Eq. (10.1), Eq. (10.2) implies that the change-of-scale affects only the *magnitude* of the intensity function $\lambda_-(x)$, and does not affect its *shape*.

From Eqs. (10.1) and (10.2) it is evident that, up to a change of magnitude of their intensity functions, the power Poisson processes $\mathcal{E}_+$ and $\mathcal{E}_-$ are invariant with respect to changes-of-scale. This observation gives rise to the following question: Are there other Poisson processes that display such scale-invariance?

The answer turns out to be almost negative. Indeed, in the context of Poisson processes over the positive half-line such scale-invariance is displayed only by Poisson processes with a power intensity: $\lambda(x) = Cx^{p-1}$ ($x > 0$), where C is a positive coefficient, and where p is a real power; the proof of this scale-invariance result is detailed in the Methods. Hence, we conclude that such scale-invariance is a characteristic feature of the power Poisson processes, $\mathcal{E}_+$ and $\mathcal{E}_-$, as well as of the harmonic Poisson process (characterized by the harmonic intensity $\lambda(x) = C/x$) [5].

10.2 Perturbation Invariance

The scale-invariance of the previous section transcends to a more general perturbation-invariance. This invariance is in the context of random multiplicative perturbations that we shall now address. This section is based on [6, 7], which study the invariance of Poisson processes with respect to general perturbation schemes.

A random multiplicative perturbation is described as follows. On the one hand, consider a random countable collection of positive points $\{x_k\}$. On the other hand, consider a general positive-valued random variable S, and generate from it a countable collection of IID copies $\{S_k\}$. The two collections are assumed independent. Multiplying each point x_k of the first collection by its corresponding copy S_k, we obtain the collection $\{S_k \cdot x_k\}$. The collection $\{S_k \cdot x_k\}$ is the random multiplicative perturbation—induced by the generic random variable S—of the initial collection $\{x_k\}$.

A result from the theory of Poisson processes (the "displacement theorem" [4]) implies that applying the random multiplicative perturbation to the power Poisson process $\mathcal{E}_+$ yields a new Poisson process $\tilde{\mathcal{E}}_+$, also with positive-valued points. Moreover, in terms of the intensity function $\lambda_+(x) = c\epsilon x^{\epsilon-1}$ of the Poisson process $\mathcal{E}_+$, the intensity function of the new Poisson process is given by

$$\tilde{\lambda}_+(x) = \mathbf{E}\left[S^{-\epsilon}\right] \cdot \lambda_+(x) \tag{10.3}$$

($x > 0$); the derivation of Eq. (10.3) is detailed in the Methods. Equation (10.3) implies that the random multiplicative perturbation affects only the *magnitude* of the intensity function $\lambda_+(x)$, and does not affect its *shape*. Equation (10.3) requires that the generic random variable S have a finite moment of order $-\epsilon$, i.e.: $\mathbf{E}\left[S^{-\epsilon}\right] < \infty$.

10.2 Perturbation Invariance

The aforementioned result from the theory of Poisson processes implies that applying the random multiplicative perturbation to the power Poisson process $\mathcal{E}_-$ yields a new Poisson process $\tilde{\mathcal{E}}_-$, also with positive-valued points. Moreover, in terms of the intensity function $\lambda_-(x) = c\epsilon x^{-\epsilon-1}$ of the Poisson process $\mathcal{E}_-$, the intensity function of the new Poisson process is given by

$$\tilde{\lambda}_-(x) = \mathbf{E}\left[S^\epsilon\right] \cdot \lambda_-(x) \tag{10.4}$$

($x > 0$); the derivation of Eq. (10.4) is detailed in the Methods. Similar to Eq. (10.3), Eq. (10.4) implies that the random multiplicative perturbation affects only the *magnitude* of the intensity function $\lambda_-(x)$, and does not affect its *shape*. Equation (10.4) requires that the generic random variable S have a finite moment of order ϵ, i.e., $\mathbf{E}[S^\epsilon] < \infty$.

Note that in case the generic random variable S is a constant—i.e., $S = s$ with probability one, where s is a positive scale—then the random multiplicative perturbation of this section reduces to the change-of-scale of the previous section. Consequently, Eq. (10.3) reduces to Eq. (10.1), and Eq. (10.4) reduces to Eq. (10.2).

From Eqs. (10.3) and (10.4) it is evident that, up to a change of magnitude of their intensity functions, the power Poisson processes, $\mathcal{E}_+$ and $\mathcal{E}_-$, are invariant with respect to random multiplicative perturbations. This observation gives rise to the following question: Are there other Poisson processes that display such perturbation-invariance?

The answer turns out to be almost negative. Indeed, in the context of Poisson processes over the positive half-line, such perturbation-invariance is displayed only by Poisson processes with a power intensity: $\lambda(x) = Cx^{p-1}$ ($x > 0$), where C is a positive coefficient, and where p is a real power; the proof of this perturbation-invariance result is detailed in the Methods. Hence, we conclude that such perturbation-invariance is a characteristic feature of the power Poisson processes, $\mathcal{E}_+$ and $\mathcal{E}_-$, as well as of the harmonic Poisson process (characterized by the harmonic intensity $\lambda(x) = C/x$) [5].

10.3 Symmetric Perturbations

In this section, we introduce the following variation to the random multiplicative perturbation of Sect. 10.2: the general random variable S is now considered real-valued and symmetric, rather than positive-valued. Namely: $S = -S$, the equality being in law.

A result from the theory of Poisson processes (the "displacement theorem" [4]) implies that applying the random multiplicative perturbation to the power Poisson process $\mathcal{E}_+$ yields a new Poisson process $\tilde{\mathcal{E}}_+$, with real-valued points. Moreover, in terms of the intensity function $\lambda_+(x) = c\epsilon x^{\epsilon-1}$ of the Poisson process $\mathcal{E}_+$, the intensity function of the new Poisson process is given by

$$\tilde{\lambda}_+ (y) = \frac{1}{2}\mathbf{E}[\,|S|^{-\epsilon}] \cdot \lambda_+ (|y|) \qquad (10.5)$$

$(-\infty < y < \infty)$; the derivation of Eq. (10.5) is detailed in the Methods. Equation (10.5) implies that the random multiplicative perturbation has the following effect on the intensity function $\lambda_+ (x)$: symmetrization, and a change of magnitude. Equation (10.5) requires that the random variable $|S|$ have a finite moment of order $-\epsilon$, i.e., $\mathbf{E}[\,|S|^{-\epsilon}] < \infty$.

The aforementioned result from the theory of Poisson processes implies that applying the random multiplicative perturbation to the power Poisson process $\mathcal{E}_-$ yields a new Poisson process $\tilde{\mathcal{E}}_-$ with real-valued points. Moreover, in terms of the intensity function $\lambda_- (x) = c\epsilon x^{-\epsilon-1}$ of the Poisson process $\mathcal{E}_-$, the intensity function of the new Poisson process is given by

$$\tilde{\lambda}_- (y) = \frac{1}{2}\mathbf{E}[\,|S|^{\epsilon}] \cdot \lambda_- (|y|) \qquad (10.6)$$

$(-\infty < y < \infty)$; the derivation of Eq. (10.6) is detailed in the Methods. Similar to Eq. (10.5), Eq. (10.6) implies that the random multiplicative perturbation has the following effect on the intensity function $\lambda_- (x)$: symmetrization, and a change of magnitude. Equation (10.6) requires that the random variable $|S|$ have a finite moment of order ϵ, i.e., $\mathbf{E}[\,|S|^{\epsilon}] < \infty$.

Equations (10.5) and (10.6) admit the structure $\tilde{\lambda}(y) = m \cdot \lambda(|y|)$, where: $\lambda(x)$ $(x > 0)$ is the intensity of the "input" Poisson process; $\tilde{\lambda}(y)$ $(-\infty < y < \infty)$ is the intensity of the "output" Poisson process; and m is a positive "magnitude constant" that depends on the generic random variable S. This observation gives rise to the following question: Are there other Poisson processes that display such a structure?

The answer turns out to be almost negative. Indeed, such a structure is displayed only by Poisson processes with a power intensity: $\lambda(x) = Cx^{p-1}$ $(x > 0)$, where C is a positive coefficient, and where p is a real power; the proof of this symmetric-perturbation result is detailed in the Methods. Hence, we conclude that such a structure is a characteristic feature of the power Poisson processes $\mathcal{E}_+$ and $\mathcal{E}_-$, as well as of the harmonic Poisson process (characterized by the harmonic intensity $\lambda(x) = C/x$) [5].

10.4 Socioeconomic Invariance

In Chap. 9, we explored the socioeconomic behavior of the power Poisson processes $\mathcal{E}_+$ and $\mathcal{E}_-$. To that end, we applied the threshold analysis of Sect. 9.3: we introduced a positive threshold level l, and then computed the disparity curves and the Lorenz curves of the power Poisson processes $\mathcal{E}_+$ and $\mathcal{E}_-$ with respect to the level l. Specifically, for the power Poisson process $\mathcal{E}_+$ we addressed the disparity curve and the Lorenz curves that correspond to the points of $\mathcal{E}_+$ that reside below the level l. And,

10.4 Socioeconomic Invariance

for the power Poisson process $\mathcal{E}_-$, we addressed the disparity curve and the Lorenz curves that correspond to the points of $\mathcal{E}_-$ that reside above the level l.

In Chap. 9, we showed that these disparity curves and Lorenz curves are invariant with respect to the threshold level l. Namely, we used the threshold level l in order to "zoom in" and to "zoom out" on parts of the power Poisson processes $\mathcal{E}_+$ and $\mathcal{E}_-$—and doing so we saw that the inherent socioeconomic equality/inequality of these parts remains the same. Hence, we obtained *socioeconomic invariance* with respect to the threshold level l.

This observation gives rise to the following question: Are there other Poisson processes, over the positive half-line, whose disparity curves and Lorenz curves display such a socioeconomic invariance with respect to the threshold level l? The answer turns out to be negative. Indeed, in [8] it was established that such a socioeconomic invariance *characterizes* the power Poisson processes $\mathcal{E}_+$ and $\mathcal{E}_-$. Hence, in particular, we conclude that: The power Lorenz curve $L_+(x) = x^{1+1/\epsilon}$ ($0 \le x \le 1$) *characterizes* the power Poisson process $\mathcal{E}_+$, and the power Lorenz curve $\bar{L}_-(x) = x^{1-1/\epsilon}$ ($0 \le x \le 1; \epsilon > 1$) *characterizes* the power Poisson process $\mathcal{E}_-$ (for exponent $\epsilon > 1$). These Lorenz-curves characterizations have Pareto perspectives which are described as follows.

The power Lorenz curve $L_+(x) = x^{1+1/\epsilon}$ ($0 \le x \le 1$) yields, via a Lorenz-curve inversion, the cumulative distribution function of the inverse-Pareto law: $F(w) = (w/l)^\epsilon$ ($0 < w < l$), with threshold level

$$l = \left(1 + \frac{1}{\epsilon}\right)\mu, \tag{10.7}$$

where μ is the corresponding mean; the derivation of this result is detailed in the Methods. Similarly, the power Lorenz curve $\bar{L}_-(x) = x^{1-1/\epsilon}$ ($0 \le x \le 1; \epsilon > 1$) yields, via a Lorenz-curve inversion, the tail distribution function of the Pareto law (for exponent $\epsilon > 1$): $\bar{F}(w) = (l/w)^\epsilon$ ($l < w < \infty$), with threshold level

$$l = \left(1 - \frac{1}{\epsilon}\right)\mu, \tag{10.8}$$

where μ is the corresponding mean; the derivation of this result is detailed in the Methods.

Interestingly, when shifting from the power Lorenz curves $L_+(x) = x^{1+1/\epsilon}$ ($0 \le x \le 1$) and $\bar{L}_-(x) = x^{1-1/\epsilon}$ ($0 \le x \le 1; \epsilon > 1$) to their corresponding distribution functions—the threshold levels of Eqs. (10.7) and (10.8) pop up, respectively. And, as the mean μ can assume any positive value, so can the threshold levels of Eqs. (10.7) and (10.8). These threshold levels are the hallmarks of the "underlying icebergs": the inverse-Pareto law is a tip of the Poisson-process iceberg $\mathcal{E}_+$, and the Pareto law is a tip of the Poisson-process iceberg $\mathcal{E}_-$. Indeed, the power Lorenz curves characterize the entire Poisson-process icebergs $\mathcal{E}_+$ and $\mathcal{E}_-$ – rather than merely their tips, the inverse-Pareto and Pareto laws.

10.5 Poor Fractality and Rich Fractality

In the previous section, we asserted that power Lorenz curves characterize the power Poisson processes $\mathcal{E}_+$ and $\mathcal{E}_-$. In this section, which is based on [9], we shall address the "poor fractality" of the power Poisson process $\mathcal{E}_+$, and the "rich fractality" of the power Poisson process $\mathcal{E}_-$. The notions of "poor fractality" and "rich fractality" are defined via Lorenz curves; as in Chap. 9, we consider a human society comprising individuals with non-negative wealth values, and denote the society's Lorenz curves $y = L(x)$ and $y = \bar{L}(x)$ ($0 \leq x, y \leq 1$).

Recall that the Lorenz curve $y = L(x)$ is defined as follows: the low (poor) $100x\%$ of the society's population are in possession of $100y\%$ of the society's overall wealth. For an arbitrary number u in the unit interval, $0 < u < 1$, let's focus on the sub-population comprising the low (poor) $100u\%$ of the society's individuals. The Lorenz curve of this sub-population is $y = L(ux)/L(u)$ ($0 \leq x, y \leq 1$).

As noted above, a fractal is invariant with respect to zooming in and zooming out. Hence, from a socioeconomic perspective, the society displays *poor fractality* [9] if the wealth distribution among the poor sub-population is identical to the wealth distribution among the entire population, i.e., their respective Lorenz curves coincide:

$$\frac{L(ux)}{L(u)} = L(x) \tag{10.9}$$

($0 \leq x \leq 1$) for all $0 < u < 1$.

In Chap. 9, we established that $L_+(x) = x^{1+1/\epsilon}$ ($0 \leq x \leq 1$) is the Lorenz curve of the power Poisson process $\mathcal{E}_+$. Evidently, the Lorenz curve $L_+(x)$ satisfies Eq. (10.9), and hence we obtain that the power Poisson process $\mathcal{E}_+$ displays poor fractality.

Recall that the Lorenz curve $y = \bar{L}(x)$ is defined as follows: the high (rich) $100x\%$ of the society's population are in possession of $100y\%$ of the society's overall wealth. For an arbitrary number u in the unit interval, $0 < u < 1$, let us focus on the sub-population comprising the high (rich) $100u\%$ of the society's individuals. The Lorenz curve of this sub-population is $y = \bar{L}(ux)/\bar{L}(u)$ ($0 \leq x, y \leq 1$).

As noted above, a fractal is invariant with respect to zooming in and zooming out. Hence, from a socioeconomic perspective, the society displays *rich fractality* [9] if the wealth distribution among the rich sub-population is identical to the wealth distribution among the entire population, i.e. their respective Lorenz curves coincide:

$$\frac{\bar{L}(ux)}{\bar{L}(u)} = \bar{L}(x) \tag{10.10}$$

($0 \leq x \leq 1$) for all $0 < u < 1$.

In Chap. 9 we established that $\bar{L}_-(x) = x^{1-1/\epsilon}$ ($0 \leq x \leq 1$) is the Lorenz curve of the power Poisson process $\mathcal{E}_-$ (for exponent $\epsilon > 1$). Evidently, the Lorenz curve $\bar{L}_-(x)$ satisfies Eq. (10.10), and hence we obtain that the power Poisson process $\mathcal{E}_-$ displays rich fractality (for exponent $\epsilon > 1$).

10.5 Poor Fractality and Rich Fractality

So, the power Poisson process $\mathcal{E}_+$ displays *poor fractality*, and the power Poisson process $\mathcal{E}_-$ displays *rich fractality*. This observation gives rise to the following question: Are there other Poisson processes that are poor-fractal and rich-fractal? The answer, as we shall now argue, is negative.

Assume that Eq. (10.9) holds. Then, the Lorenz curve $y = L(x)$ is a power function: $L(x) = x^p$ ($0 \leq x \leq 1$). Moreover, as the Lorenz curve $y = L(x)$ is a convex function, the power p is restricted to the range $p \geq 1$. Hence, we obtain that poor fractality implies the Lorenz curve $L_+(x) = x^{1+1/\epsilon}$ ($0 \leq x \leq 1$), where ϵ is a positive exponent. In Sect. 10.4, we established that the Lorenz curve $L_+(x) = x^{1+1/\epsilon}$ characterizes—in the context of Poisson processes over the positive half-line—the power Poisson process $\mathcal{E}_+$. Thus, we conclude that: poor fractality, defined by Eq. (10.9), *characterizes* the Poisson processes $\mathcal{E}_+$.

Assume that Eq. (10.10) holds. Then, the Lorenz curve $y = \bar{L}(x)$ is a power function: $\bar{L}(x) = x^p$ ($0 \leq x \leq 1$). Moreover, as the Lorenz curve $y = \bar{L}(x)$ is a concave function, the power p is restricted to the range $0 < p < 1$. Hence, we obtain that rich fractality implies the Lorenz curve $\bar{L}_-(x) = x^{1-1/\epsilon}$ ($0 \leq x \leq 1$), where ϵ is an exponent taking values in the range $\epsilon > 1$. In Sect. 10.4, we established that the Lorenz curve $\bar{L}_-(x) = x^{1-1/\epsilon}$ characterizes—in the context of Poisson processes over the positive half-line—the power Poisson process $\mathcal{E}_-$ (with exponent $\epsilon > 1$). Thus, we conclude that: rich fractality, defined by Eq. (10.10), *characterizes* the Poisson processes $\mathcal{E}_-$ (with exponent $\epsilon > 1$).

10.6 Renormalization

Often, fractals admit the following structure [1–3]: the whole fractal is the union of several scaled-down copies of itself. For example [1–3]: (i) the Cantor set is the union of two copies of itself, where each copy is shrunk by the scale $1/3$; (ii) the Koch curve is the union of four copies of itself, where each copy is shrunk by the scale $1/3$; (iii) the Sierpinski triangle is the union of three copies of itself, where each copy is shrunk by the scale $1/2$. This fractal structure is described, mathematically, by invariance with respect to linear renormalization schemes [10].

In the context of Poisson processes over the positive half-line, the linear renormalization scheme is formulated as follows [11]. Initiating from a general Poisson process $\mathcal{P}$: generate n IID copies of the Poisson process $\mathcal{P}$, take the union of the n IID copies, and then apply the change-of-scale $x \mapsto s \cdot x$ (where s is a positive scale) to the points of the union. The collection of the resulting points constitute the renormalization of the Poisson process $\mathcal{P}$.

Denote by $\lambda(x)$ ($x > 0$) the intensity function of the initial Poisson process $\mathcal{P}$. A result from the theory of Poisson processes (the "superposition theorem" [4]) implies that the union is a Poisson process over the positive half-line, and that its intensity function is $n \cdot \lambda(x)$ ($x > 0$). In turn, the renormalization of the Poisson process $\mathcal{P}$ is also a Poisson process over the positive half-line, and its intensity function is given by

$$\tilde{\lambda}(x) = \frac{n}{s}\lambda\left(\frac{x}{s}\right) \qquad (10.11)$$

($x > 0$); the derivation of Eq. (10.11) is detailed in the Methods.

Consider the power Poisson processes, $\mathcal{E}_+$ and $\mathcal{E}_-$, with their respective intensity functions $\lambda_+(x) = c\epsilon x^{\epsilon-1}$ and $\lambda_-(x) = c\epsilon x^{-\epsilon-1}$. Initiating from the power Poisson process $\mathcal{E}_+$, and setting the scale $s = n^{1/\epsilon}$, implies that $\tilde{\lambda}(x) = \lambda_+(x)$. Similarly, initiating from the power Poisson process $\mathcal{E}_-$, and setting the scale $s = n^{-1/\epsilon}$, implies that $\tilde{\lambda}(x) = \lambda_-(x)$. Hence, when a proper scaling is applied, the intensity functions $\lambda_+(x)$ and $\lambda_-(x)$ are *fixed-points* of Eq. (10.11).

In general, an intensity function $\lambda(x)$ is a fixed-point of Eq. (10.11) if the following holds: to any given number of copies n corresponds a scale s_n such that $\tilde{\lambda}(x) = \lambda(x)$ ($n = 1, 2, 3, \ldots$). Moreover, the parameter n can be extended from the integer range $n = 1, 2, 3, \ldots$ to the continuous range $n > 0$. The natural question that arises is: What are *all* the fixed-points of Eq. (10.11)?

Observing Eq. (10.11) it is straightforward to deduce that all the fixed-point intensity functions are characterized by the following power form: $\lambda(x) = Cx^{p-1}$ ($x > 0$), where C is a positive coefficient, and where $p \neq 0$ is a real power. We emphasize that the harmonic intensity $\lambda(x) = C/x$ is *not* a fixed-point of Eq. (10.11).[1] Thus, we conclude that "renormalization fractality"—defined via the fixed-points of Eq. (10.11)—*characterizes* the power Poisson processes $\mathcal{E}_+$ and $\mathcal{E}_-$.

10.7 Summary

In this chapter, we explored various fractal features of the power Poisson processes $\mathcal{E}_+$ and $\mathcal{E}_-$. The first fractal feature was scale-invariance: we investigated the effect of change-of-scale transformations on the points of the power Poisson processes $\mathcal{E}_+$ and $\mathcal{E}_-$, and established that these processes are invariant with respect to such transformations; we further established that this scale-invariance characterizes the power Poisson processes, $\mathcal{E}_+$ and $\mathcal{E}_-$, as well as the harmonic Poisson process. The second fractal feature was perturbation-invariance: we investigated the effect of random multiplicative perturbations on the points of the power Poisson processes, $\mathcal{E}_+$ and $\mathcal{E}_-$, and established that these processes are invariant with respect to such perturbations; we further established that this perturbation-invariance characterizes the power Poisson processes, $\mathcal{E}_+$ and $\mathcal{E}_-$, as well as the harmonic Poisson process.

The third fractal feature was socioeconomic invariance: we asserted that the invariances described in Chap. 9—of the disparity curves and the Lorenz curves of the power Poisson processes, $\mathcal{E}_+$ and $\mathcal{E}_-$, with respect to the threshold level l—characterize these processes. The fourth fractal feature was poor/rich fractality: using a Lorenz-curves perspective, we showed that poor fractality and rich fractality characterize, respectively, the power Poisson processes $\mathcal{E}_+$ and $\mathcal{E}_-$. The fifth fractal feature

---

[1] The harmonic Poisson process (characterized by the harmonic intensity $\lambda(x) = C/x$) turns out to be fractal in the context of a different, *nonlinear*, renormalization scheme [5].

was renormalization: we investigated the effect of linear renormalization schemes on the power Poisson processes, $\mathcal{E}_+$ and $\mathcal{E}_-$, and established that these processes—and only these processes—are the fixed-points of such renormalizations.

10.8 Methods

10.8.1 Equations (10.1), (10.2), and (10.11)

Consider a Poisson process $\mathcal{P}$ over a real domain $\mathcal{X}$, with intensity function $\lambda_\mathcal{P}(x)$ ($x \in \mathcal{X}$). Further consider a monotone function $y = f(x)$ that maps the real domain $\mathcal{X}$ to another real domain $\mathcal{Y}$. Applying the map $x \mapsto y = f(x)$ to each and every point of the Poisson process $\mathcal{P}$, we obtain the collection of points

$$\mathcal{Q} = \{f(x)\}_{x \in \mathcal{P}}. \tag{10.12}$$

The "displacement theorem" of the theory of Poisson processes [4] asserts that $\mathcal{Q}$ is a Poisson process over the real domain $\mathcal{Y}$, with intensity function

$$\lambda_\mathcal{Q}(y) = \frac{\lambda_\mathcal{P}\left[f^{-1}(y)\right]}{\left|f'\left[f^{-1}(y)\right]\right|} \tag{10.13}$$

($y \in \mathcal{Y}$).

Now, set the domain $\mathcal{X}$ to be the positive half-line, and set $f(x) = s \cdot x$ ($x > 0$). For this setting the domain $\mathcal{Y}$ is the positive half-line, and Eq. (10.13) yields

$$\lambda_\mathcal{Q}(y) = \frac{1}{s}\lambda_\mathcal{P}\left(\frac{y}{s}\right) \tag{10.14}$$

($y > 0$). Consider the following power intensity: $\lambda_\mathcal{P}(x) = Cx^{p-1}$ ($x > 0$), where C is a positive coefficient and where p is a real power. Applying Eq. (10.14) to this power intensity yields

$$\lambda_\mathcal{Q}(y) = \frac{1}{s}C\left(\frac{y}{s}\right)^{p-1} = Cy^{p-1}s^{-p} = s^{-p} \cdot \lambda_\mathcal{P}(y) \tag{10.15}$$

($y > 0$). Hence, we obtain that: (i) $\mathcal{P} = \mathcal{E}_+ \Rightarrow \lambda_\mathcal{P}(x) = \lambda_+(x) \Rightarrow \lambda_\mathcal{Q}(y) = s^{-\epsilon} \cdot \lambda_+(y)$; and (ii) $\mathcal{P} = \mathcal{E}_- \Rightarrow \lambda_\mathcal{P}(x) = \lambda_-(x) \Rightarrow \lambda_\mathcal{Q}(y) = s^\epsilon \cdot \lambda_-(y)$. This proves Eqs. (10.1) and (10.2). We also obtain that: $\lambda_\mathcal{P}(x) = n \cdot \lambda(x) \Rightarrow \lambda_\mathcal{Q}(y) = (n/s)\lambda_\mathcal{P}(y/s)$; this proves Eq. (10.11)

Assume now that the change-of-scale, $x \mapsto s \cdot x$, results only in a change of magnitude of the intensity function. Namely:

$$\lambda_{\mathcal{Q}}(y) = \phi(s) \cdot \lambda_{\mathcal{P}}(y) \tag{10.16}$$

($y > 0$), where $\phi(s)$ is a positive-valued function. Combining together Eqs. (10.14) and (10.16) yields $\frac{\lambda_{\mathcal{P}}(y/s)}{\lambda_{\mathcal{P}}(y)} = s\phi(s)$ ($y, s > 0$). Hence, we obtain that the intensity function must be a power-law: $\lambda_{\mathcal{P}}(x) = Cx^{p-1}$ ($x > 0$), where C is a positive coefficient and where p is a real power. In turn, Eq. (10.15) implies that the change of magnitude is given by: $\phi(s) = s^{-p}$.

10.8.2 Equations (10.3) and (10.4)

Consider a Poisson process $\mathcal{P}$ over the positive half-line, with intensity function $\lambda_{\mathcal{P}}(x)$ ($x > 0$), and denote by $\{x_k\}$ the points of $\mathcal{P}$. Consider a general positive-valued random variable S, with probability density function $g(s)$ ($s > 0$), and generate from it a countable collection of IID copies $\{S_k\}$ that are independent of $\mathcal{P}$. Introduce the collection of points $\mathcal{Q} = \{S_k \cdot x_k\}$. The "displacement theorem" of the theory of Poisson processes [4] asserts that $\mathcal{Q}$ is a Poisson process over the positive half-line, with intensity function

$$\lambda_{\mathcal{Q}}(y) = \int_0^\infty \lambda_{\mathcal{P}}\left(\frac{y}{s}\right) \frac{g(s)}{s} ds \tag{10.17}$$

($y > 0$). Consider the following power intensity: $\lambda_{\mathcal{P}}(x) = Cx^{p-1}$ ($x > 0$), where C is a positive coefficient and where p is a real power. Applying Eq. (10.17) to this power intensity yields

$$\begin{aligned}\lambda_{\mathcal{Q}}(y) &= \int_0^\infty C\left(\frac{y}{s}\right)^{p-1} \frac{g(s)}{s} ds \\ &= Cy^{p-1} \int_0^\infty s^{-p} g(s) ds \\ &= \mathbf{E}\left[S^{-p}\right] \cdot \lambda_{\mathcal{P}}(y)\end{aligned} \tag{10.18}$$

($y > 0$). Hence, we obtain that: (i) $\mathcal{P} = \mathcal{E}_+ \Rightarrow \lambda_{\mathcal{P}}(x) = \lambda_+(x) \Rightarrow \lambda_{\mathcal{Q}}(y) = \mathbf{E}\left[S^{-\epsilon}\right] \cdot \lambda_+(y)$; and (ii) $\mathcal{P} = \mathcal{E}_- \Rightarrow \lambda_{\mathcal{P}}(x) = \lambda_-(x) \Rightarrow \lambda_{\mathcal{Q}}(y) = \mathbf{E}\left[S^{\epsilon}\right] \cdot \lambda_-(y)$. This proves Eqs. (10.3) and (10.4).

Assume now that the transformation $\mathcal{P} \mapsto \mathcal{Q}$ results only in a change of magnitude of the intensity function. Namely:

$$\lambda_{\mathcal{Q}}(y) = m \cdot \lambda_{\mathcal{P}}(y) \tag{10.19}$$

($y > 0$), where m is a positive "magnitude constant" that depends on the generic random variable S. Combining together Eqs. (10.17) and (10.19) yields

10.8 Methods

$$m \cdot \lambda_\mathcal{P}(y) = \int_0^\infty \lambda_\mathcal{P}\left(\frac{y}{s}\right) \frac{g(s)}{s} ds \quad (10.20)$$

($y > 0$). Equation (10.20) can hold if and only if the intensity function is a power: $\lambda_\mathcal{P}(x) = Cx^{p-1}$ ($x > 0$), where C is a positive coefficient and where p is a real power. In turn, Eqs. (10.18) implies that $m = \mathbf{E}\left[S^{-p}\right]$.

10.8.3 Equations (10.5) and (10.6)

Consider the setting of the previous subsection, but now with the random variable S being real-valued and symmetric (rather than positive-valued). In this case Eq. (10.17) becomes

$$\lambda_\mathcal{Q}(y) = \int_0^\infty \lambda_\mathcal{P}\left(\frac{|y|}{s}\right) \frac{g(s)}{s} ds \quad (10.21)$$

($-\infty < y < \infty$). In turn, Eq. (10.18) becomes

$$\lambda_\mathcal{Q}(y) = C |y|^{p-1} \int_0^\infty s^{-p} g(s) ds$$
$$= \tfrac{1}{2}\mathbf{E}[|S|^{-p}] \cdot \lambda_\mathcal{P}(|y|) \quad (10.22)$$

($-\infty < y < \infty$). Hence, we obtain that: (i) $\mathcal{P} = \mathcal{E}_+ \Rightarrow \lambda_\mathcal{P}(x) = \lambda_+(x)$ $\Rightarrow \lambda_\mathcal{Q}(y) = \tfrac{1}{2}\mathbf{E}\left[S^{-\epsilon}\right] \cdot \lambda_+(|y|)$; and (ii) $\mathcal{P} = \mathcal{E}_- \Rightarrow \lambda_\mathcal{P}(x) = \lambda_-(x) \Rightarrow \lambda_\mathcal{Q}(y) = \tfrac{1}{2}\mathbf{E}\left[S^\epsilon\right] \cdot \lambda_-(|y|)$. This proves Eqs. (10.5) and (10.6).

Assume now that the transformation $\mathcal{P} \mapsto \mathcal{Q}$ results in the following symmetrization:

$$\lambda_\mathcal{Q}(y) = m \cdot \lambda_\mathcal{P}(|y|) \quad (10.23)$$

($-\infty < y < \infty$), where m is a positive "magnitude constant" that depends on the generic random variable S. Combining together Eqs. (10.21) and (10.23) yields

$$m \cdot \lambda_\mathcal{P}(|y|) = \int_0^\infty \lambda_\mathcal{P}\left(\frac{|y|}{s}\right) \frac{g(s)}{s} ds \quad (10.24)$$

($-\infty < y < \infty$), and Eq. (10.24) is identical to Eq. (10.20). Hence, Eq. (10.24) can hold if and only if the intensity function is a power: $\lambda_\mathcal{P}(x) = Cx^{p-1}$ ($x > 0$), where C is a positive coefficient and where p is a real power. In turn, Eqs. (10.22) implies that $m = \tfrac{1}{2}\mathbf{E}[|S|^{-p}]$.

10.8.4 Equations (10.7) and (10.8)

Consider the power Lorenz curve $L_+(x) = x^{1+1/\epsilon}$ ($0 \leq x \leq 1$). Plugging this Lorenz curve into the left-hand side of Eq. (9.7) of Chap. 9 yields

$$x^{1+1/\epsilon} = \frac{1}{\mu} \int_0^x F^{-1}(u) \, du \tag{10.25}$$

($0 \leq x \leq 1$), where μ is the corresponding mean. In turn, differentiating both sides of Eq. (10.25) yields

$$\left(1 + \frac{1}{\epsilon}\right) x^{1/\epsilon} = \frac{1}{\mu} F^{-1}(x) \tag{10.26}$$

($0 \leq x \leq 1$). Substituting $x = F(w)$ further yields

$$\left(1 + \frac{1}{\epsilon}\right) F(w)^{1/\epsilon} = \frac{1}{\mu} w. \tag{10.27}$$

Setting

$$l = \left(1 + \frac{1}{\epsilon}\right) \mu, \tag{10.28}$$

we obtain from Eq. (10.27) the following inverse-Pareto cumulative distribution function:

$$F(w) = \left(\frac{w}{l}\right)^\epsilon \tag{10.29}$$

($0 < w < l$).

Consider the power Lorenz curve $\bar{L}_-(x) = x^{1-1/\epsilon}$ ($0 \leq x \leq 1$), with exponent $\epsilon > 1$. Plugging this Lorenz curve into the left-hand side of Eq. (9.8) of Chap. 9 yields

$$x^{1-1/\epsilon} = \frac{1}{\mu} \int_0^x \bar{F}^{-1}(u) \, du \tag{10.30}$$

($0 \leq x \leq 1$), where μ is the corresponding mean. In turn, differentiating both sides of Eq. (10.30) yields

$$\left(1 - \frac{1}{\epsilon}\right) x^{-1/\epsilon} = \frac{1}{\mu} \bar{F}^{-1}(x) \tag{10.31}$$

($0 \leq x \leq 1$). Substituting $x = \bar{F}(w)$ further yields

$$\left(1 - \frac{1}{\epsilon}\right) \bar{F}(w)^{-1/\epsilon} = \frac{1}{\mu} w. \tag{10.32}$$

Setting

10.8 Methods

$$l = \left(1 - \frac{1}{\epsilon}\right)\mu, \tag{10.33}$$

we obtain from Eq. (10.32) the following Pareto tail distribution function:

$$\bar{F}(w) = \left(\frac{l}{w}\right)^{\epsilon} \tag{10.34}$$

($l < w < \infty$).

References

1. B.B. Mandelbrot, *The Fractal Geometry of Nature* (Freeman, New York, 1983)
2. M. Schroeder, *Fractals, Chaos, Power Laws: Minutes from an Infinite Paradise* (Dover, Mineola, 2009)
3. K. Falconer, *Fractal Geometry* (Wiley, Hoboken, 2014)
4. J.F.C. Kingman, *Poisson Processes* (Oxford University Press, Oxford, 1993)
5. I. Eliazar, Ann. Phys. **380**, 168 (2017)
6. I. Eliazar, J. Klafter, Phys. Rev. E **79**, 011103 (2009)
7. I. Eliazar, J. Phys. A: Math. Theor. **45**, 415103 (2012)
8. I. Eliazar, Europhys. Lett. **124**, 50003 (2018)
9. I. Eliazar, M.H. Cohen, Phys. A **402**, 30 (2014)
10. M.F. Barnsley, *Fractals Everywhere* (Dover, Mineola, 2012)
11. I. Eliazar, Phys. Rev. E **87**, 052125 (2013)

Chapter 11
Sums

As the order statistics of the power Poisson process $\mathcal{E}_+$ diverge to infinity, their sum diverges. On the other hand, as the order statistics of the power Poisson process $\mathcal{E}_-$ converge to zero, their sum may converge. In this chapter, we explore sums induced by the power Poisson process $\mathcal{E}_-$. As we shall see, this sum analysis will give rise to two Lévy-stable laws: one-sided and symmetric.

11.1 One-Sided Lévy Law I

Consider the power Poisson process $\mathcal{E}_-$, with its intensity function $\lambda_-(x) = c\epsilon x^{-\epsilon-1}$ ($x > 0$). In this section, we focus on the sum of the points of this process:

$$\Sigma_1 = \sum_{x \in \mathcal{E}_-} x. \tag{11.1}$$

A general result from the theory of Poisson processes (the "characteristic functional" [1]) implies that the the sum Σ_1 is convergent if and only if the intensity's exponent is in the range $0 < \epsilon < 1$.

In the exponent range $0 < \epsilon < 1$, the sum Σ_1 is a positive-valued random variable, and its Laplace transform is

$$\mathbf{E}\left[\exp(-\theta \Sigma_1)\right] = \exp\left[-\Gamma(1-\epsilon) c \cdot \theta^\epsilon\right] \tag{11.2}$$

($\theta \geq 0$); the derivation of Eq. (11.2) is detailed in the Methods. Differentiating both sides of Eq. (11.2) with respect to the Laplace variable θ, and then setting $\theta = 0$, implies that the sum Σ_1 has a divergent mean: $\mathbf{E}[\Sigma_1] = \infty$.

Combined together with a matching Tauberian theorem [2], Eq. (11.2) implies that the tail distribution function of the sum Σ_1 is asymptotically equivalent to the power function $\Lambda_-(l) = cl^{-\epsilon}$—the mean number of points of the power Poisson process $\mathcal{E}_-$ that reside above the threshold l. Specifically:

$$\Pr(\Sigma_1 > l) \approx \Lambda_-(l) \qquad (11.3)$$

(as $l \to \infty$); the derivation of Eq. (11.3) is detailed in the Methods.

The Laplace transform of Eq. (11.2) characterizes the *one-sided Lévy law*, which arises via the generalized Central Limit Theorem (in the context of sums of positive-valued IID random variables with infinite means) [3, 4]. Albeit for the exponent $\epsilon = \frac{1}{2}$, there is no simple "closed-form" formula for the probability density function of the one-sided Lévy law [5, 6]. The one-sided Lévy law with exponent $\epsilon = \frac{1}{2}$ appears in the context of the *first passage times* of Brownian motion [7, 8].

11.2 Symmetric Lévy Law I

Consider the power Poisson process $\mathcal{E}_-$, with its intensity function $\lambda_-(x) = c\epsilon x^{-\epsilon-1}$ ($x > 0$). Applying the setting of Sect. 10.3 of Chap. 10, we attach a random sign to each and every point of the power Poisson process $\mathcal{E}_-$. Specifically, in terms of this setting: the random variable S admits the value -1 with probability $\frac{1}{2}$, and admits the value $+1$ with probability $\frac{1}{2}$. Equation (10.6) of Chap. 10 implies that the attachment of random signs transforms the power Poisson process $\mathcal{E}_-$ to a new Poisson process $\tilde{\mathcal{E}}_-$, with real-valued points and with the intensity function $\tilde{\lambda}_-(y) = \frac{1}{2}\lambda_-(|y|)$ ($-\infty < y < \infty$).

In this section, we focus on the sum of the points of the power Poisson process $\tilde{\mathcal{E}}_-$:

$$\Sigma_2 = \sum_{y \in \tilde{\mathcal{E}}_-} y. \qquad (11.4)$$

A general result from the theory of Poisson processes (the "characteristic functional" [1]) implies that the the sum Σ_2 is convergent if and only if the intensity's exponent is in the range $0 < \epsilon < 2$.

In the exponent range $0 < \epsilon < 2$ the sum Σ_2 is a real-valued and symmetric random variable (i.e., $\Sigma_2 = -\Sigma_2$, the equality being in law), and its Fourier transform is

$$\mathbf{E}\left[\exp(i\theta\Sigma_2)\right] = \exp\left[-\frac{\pi}{\Gamma(\epsilon)\sin\left(\frac{1}{2}\pi\epsilon\right)} c \cdot |\theta|^\epsilon\right] \qquad (11.5)$$

($-\infty < \theta < \infty$); the derivation of Eq. (11.5) is detailed in the Methods. Differentiating both sides of Eq. (11.5) twice with respect to the Fourier variable θ, and then setting $\theta = 0$, implies that the sum Σ_2 has a divergent second moment: $\mathbf{E}[\Sigma_2^2] = \infty$. Consequently, the sum Σ_2 has a divergent variance: $\mathbf{Var}[\Sigma_2] = \infty$.

11.2 Symmetric Lévy Law I

Combined together with a matching Tauberian theorem [2], Eq. (11.5) implies that the tail distribution function of the sum Σ_2 is asymptotically equivalent to the power function $\Lambda_-(l) = c l^{-\epsilon}$—the mean number of points of the power Poisson process $\mathcal{E}_-$ that reside above the threshold l. Specifically:

$$\Pr(\Sigma_2 > l) \approx \Lambda_-(l) \tag{11.6}$$

(as $l \to \infty$); the derivation of Eq. (11.6) is detailed in the Methods.

The Fourier transform of Eq. (11.5) characterizes the *symmetric Lévy law*, which arises via the generalized Central Limit Theorem (in the context of sums of real-valued IID random variables with infinite variances) [3, 4]. Albeit for the exponent $\epsilon = 1$, there is no simple "closed-form" formula for the probability density function of the symmetric Lévy law [5, 6]. The symmetric Lévy law with exponent $\epsilon = 1$ is the *Cauchy law* [9, 10].

11.3 Uniform Random Scattering

Consider a uniform random scattering of points over the d-dimensional Euclidean space $\mathbb{R}^d$. This scattering is facilitated by a Poisson process $\mathcal{P}$ with a constant intensity function over $\mathbb{R}^d$. Distances in $\mathbb{R}^d$ are measured via a general norm $\|\cdot\|$; in particular, $\|\cdot\|$ can be the standard Euclidean norm. In this section, we set our focus on the following collection:

$$\left\{ \frac{1}{\|\mathbf{x}\|^q} \right\}_{\mathbf{x} \in \mathcal{P}}, \tag{11.7}$$

where q is a positive power.

Evidently, the collection of Eq. (11.7) comprises positive values. A general result from the theory of Poisson processes (the "displacement theorem" [1]) implies that this collection is a Poisson process over the positive half-line. We shall now argue that, in effect, the collection of Eq. (11.7) is the power Poisson process $\mathcal{E}_-$.

Set a positive level l. A value $1/\|\mathbf{x}\|^q$ of the collection of Eq. (11.7) is greater than the level l if and only if the corresponding point $\mathbf{x}$ in $\mathbb{R}^d$ is within a distance $1/l^{1/q}$ of the origin. Hence, the number of values of the collection of Eq. (11.7) that are greater than the level l is equal to the number of points of the Poisson process $\mathcal{P}$ that are within a distance $1/l^{1/q}$ of the origin. The mean number of these points is given by the integral of the intensity function over a ball of radius $r = 1/l^{1/q}$ that is centered at the origin; the ball and its radius are with respect to the general norm $\|\cdot\|$. As the intensity function is constant, we obtain that this mean number is given by

$$c_* \cdot v_d \cdot r^d = c_* \cdot v_d \cdot l^{-d/q}, \tag{11.8}$$

where c_* is the value of the constant intensity function, and v_d is the volume of the $\mathbb{R}^d$ unit ball (with respect to the general norm $\|\cdot\|$).

As established in Chap. 4, the power Poisson process $\mathcal{E}_-$ is characterized by its power function $\Lambda_-(l) = cl^{-\epsilon}$—the mean number of points of the power Poisson process $\mathcal{E}_-$ that reside above the threshold l. The right-hand side of Eq. (11.8) coincides with the power function $\Lambda_-(l) = cl^{-\epsilon}$, where the coefficient is $c = c_* \cdot v_d$, and the exponent is $\epsilon = d/q$. Hence, we conclude that: The collection of Eq. (11.7) is equal, in law, to the power Poisson process $\mathcal{E}_-$.

11.4 One-Sided Lévy Law II

We begin this section with a motivating example: point-sources of radiation in the three-dimensional Euclidean space $\mathbb{R}^3$, and a detector at the origin of $\mathbb{R}^3$. The point-sources are labeled by the index k, the position of point-source k is $\mathbf{x}_k$ (a vector in $\mathbb{R}^3$), and the magnitude of point-source k is M_k (a positive value). The effect of radiation is inversely proportional to the square of the Euclidean distance. Hence, denoting by $\|\cdot\|_2$ the Euclidean norm, the effect of point-source k on the detector is $M_k / \|\mathbf{x}_k\|_2^2$. In turn, the *aggregate effect* of all point-sources on the detector is $\sum_k M_k / \|\mathbf{x}_k\|_2^2$.

Motivated by the example of radiation, we shift from the three-dimensional Euclidean space $\mathbb{R}^3$ to the d-dimensional Euclidean space $\mathbb{R}^d$, and from the Euclidean norm $\|\cdot\|_2$ to a general norm $\|\cdot\|$. The point-sources and the detector are as in the example of radiation, yet now the effect of "radiation" is considered inversely proportional to the qth power of the distance (with respect to the general norm $\|\cdot\|$). Hence, the aggregate effect of all point-sources on the detector is given by

$$A_1 = \sum_k \frac{M_k}{\|\mathbf{x}_k\|^q} . \tag{11.9}$$

We introduce three assumptions regarding the statistics of the point-sources. First, we consider the positions $\{\mathbf{x}_k\}$ of the point-sources to be scattered randomly and uniformly over $\mathbb{R}^d$, as described in Sect. 11.3. Second, we consider the magnitudes $\{M_k\}$ of the point-sources to be positive-valued IID random variables with finite moments. Third, we consider the positions $\{\mathbf{x}_k\}$ and the magnitudes $\{M_k\}$ to be independent of each other.

The conclusion of Sect. 11.3 implies that the collection $\{1/\|\mathbf{x}_k\|^q\}$ is equal, in law, to the power Poisson process $\mathcal{E}_-$ with exponent $\epsilon = d/q$. In turn, Eq. (10.4) of Chap. 10 implies that the collection $\{M_k/\|\mathbf{x}_k\|^q\}$ is also equal, in law, to the power Poisson process $\mathcal{E}_-$ with exponent $\epsilon = d/q$. Hence, provided that $q > d$, we conclude that: The aggregate A_1 is equal, in law, to the *one-sided Lévy* sum Σ_1 of Sect. 11.1 with exponent $\epsilon = d/q$. We emphasize that the aggregate A_1 is essentially invariant with respect to the magnitudes $\{M_k\}$.

11.5 Symmetric Lévy Law II

We begin this section with a motivating example: point-masses in the three-dimensional Euclidean space $\mathbb{R}^3$, and a unit-mass at the origin of $\mathbb{R}^3$. The point-masses are labeled by the index k, the position of point-mass k is $\mathbf{x}_k$ (a vector in $\mathbb{R}^3$), and the mass of point-mass k is M_k (a positive value). The gravitational pull between two masses is inversely proportional to the square of the Euclidean distance between them. Hence, denoting by $\|\cdot\|_2$ the Euclidean norm, the pull that point-mass k exerts on the unit-mass is $M_k/\|\mathbf{x}_k\|_2^2$; this pull is exerted in the direction $\mathbf{u}_k = \mathbf{x}_k/\|\mathbf{x}_k\|_2$. In turn, the aggregate pull exerted on the unit-mass is $\sum_k M_k \mathbf{u}_k/\|\mathbf{x}_k\|_2^2$. Finally, consider the pull exerted on the unit-mass along the direction $\mathbf{u}$, an arbitrary vector in $\mathbb{R}^3$ with unit norm. Denoting by $\langle \cdot, \cdot \rangle_2$ the Euclidean scalar product, we obtain that the pull along the direction $\mathbf{u}$ is given by: $\sum_k S_k/\|\mathbf{x}_k\|_2^2$, where $S_k = M_k \cdot \langle \mathbf{u}_k, \mathbf{u} \rangle_2$.

Motivated by the example of gravitation, we shift from the three-dimensional Euclidean space $\mathbb{R}^3$ to the d-dimensional Euclidean space $\mathbb{R}^d$, and from the Euclidean norm $\|\cdot\|_2$ to a general norm $\|\cdot\|$. The point-masses and the unit-mass are as in the example of gravitation, yet now the "gravitational pull" is considered inversely proportional to the qth power of the distance (with respect to the general norm $\|\cdot\|$). Hence, the aggregate 'pull' exerted on the unit-mass along an arbitrary direction is given by

$$A_2 = \sum_k \frac{S_k}{\|\mathbf{x}_k\|^q} \cdot \qquad (11.10)$$

We introduce three assumptions regarding the statistics of the point-masses. First, we consider the positions $\{\mathbf{x}_k\}$ of the point-masses to be scattered randomly and uniformly over $\mathbb{R}^d$, as described in Sect. 11.3. Second, we consider the scalars $\{S_k\}$ to be real-valued and symmetric IID random variables with finite absolute moments. Third, we consider the positions $\{\mathbf{x}_k\}$ and the scalars $\{S_k\}$ to be independent of each other.

The conclusion of Sect. 11.3 implies that the collection $\{1/\|\mathbf{x}_k\|^q\}$ is equal, in law, to the power Poisson process $\mathcal{E}_-$ with exponent $\epsilon = d/q$. In turn, Eq. (10.6) of Chap. 10 implies that the collection $\{S_k/\|\mathbf{x}_k\|^q\}$ is equal, in law, to the power Poisson process $\tilde{\mathcal{E}}_-$ with exponent $\epsilon = d/q$. Hence, provided that $2q > d$ we conclude that: The aggregate A_2 is equal, in law, to the *symmetric Lévy* sum Σ_2 of Sect. 11.2 with exponent $\epsilon = d/q$. We emphasize that the aggregate A_2 is essentially invariant with respect to the scalars $\{S_k\}$.

The motivating example of gravitation pull—with dimension $d = 3$, Euclidean norm, and power $q = 2$—yields the *Holtsmark law* [11, 12]. Namely, the Holtsmark law is the symmetric Lévy law with exponent $\epsilon = 3/2$.

11.6 Summary

In this chapter, we explored sums induced by the power Poisson process $\mathcal{E}_-$. First, we addressed the sum of the points of the power Poisson process $\mathcal{E}_-$: we showed that this sum is convergent if and only if the exponent ϵ is in the range $0 < \epsilon < 1$—in which case the sum is governed by the one-sided Lévy law. Second, we attached random signs to the points of the power Poisson process $\mathcal{E}_-$, and then addressed their sum: we showed that this sum is convergent if and only if the exponent ϵ is in the range $0 < \epsilon < 2$—in which case the sum is governed by the symmetric Lévy law. Armed with these results, we further showed how the one-sided and the symmetric Lévy laws emerge from the aggregate effect of points that are scattered randomly and uniformly over the d-dimensional Euclidean space.

11.7 Methods

Consider a Poisson process $\mathcal{P}$ over a real domain $\mathcal{D}$, with intensity function $\lambda(x)$ ($x \in \mathcal{D}$). Set $\phi(x)$ ($x \in \mathcal{D}$) to be a general, complex-valued, test function that is defined over the domain $\mathcal{D}$. The characteristic functional of the Poisson process $\mathcal{P}$ with respect to the test function $\phi(x)$ is given by [1]:

$$\mathbf{E}\left[\prod_{x \in \mathcal{P}} \phi(x)\right] = \exp\left\{-\int_{\mathcal{D}} [1 - \phi(x)] \lambda(x) \, dx\right\}. \quad (11.11)$$

Equation (11.11) holds valid for all test functions $\phi(x)$ for which the integral appearing on the right-hand side of Eq. (11.11) is convergent. We will use Eq. (11.11) in the derivations of Eqs. (11.2) and (11.5).

11.7.1 Equations (11.2) and (11.3)

Consider the power Poisson process $\mathcal{E}_-$, over the positive half-line, with its intensity function $\lambda_-(x) = c\epsilon x^{-\epsilon-1}$ ($x > 0$). Set the test function of Eq. (11.11) to be $\phi(x) = \exp(-\theta x)$ ($x > 0$), where θ is a non-negative parameter. Equation (11.1) implies that

$$\mathbf{E}\left[\exp(-\theta \Sigma_1)\right] = \mathbf{E}\left[\exp\left(-\theta \sum_{x \in \mathcal{E}_-} x\right)\right]$$
$$= \mathbf{E}\left[\prod_{x \in \mathcal{E}_-} \exp(-\theta x)\right] = \mathbf{E}\left[\prod_{x \in \mathcal{E}_-} \phi(x)\right]. \quad (11.12)$$

On the other hand, we have

11.7 Methods

$$\int_{\mathcal{D}} [1 - \phi(x)] \lambda(x) \, dx \tag{11.13}$$

(setting $\mathcal{D} = (0, \infty)$, $\phi(x) = \exp(-\theta x)$, and $\lambda(x) = \lambda_-(x)$)

$$= \int_0^\infty [1 - \exp(-\theta x)] c\epsilon x^{-\epsilon-1} \, dx \tag{11.14}$$

(applying integration-by-parts and a change-of-variables)

$$= c\theta^\epsilon \int_0^\infty \exp(-u) u^{-\epsilon} \, du \,. \tag{11.15}$$

Using the gamma function, Eqs. (11.13)–(11.15) yield

$$\int_{\mathcal{D}} [1 - \phi(x)] \lambda(x) \, dx = c\theta^\epsilon \cdot \Gamma(1 - \epsilon) \,. \tag{11.16}$$

The integral of Eq. (11.16) is convergent if and only if the exponent ϵ is in the range $0 < \epsilon < 1$. For such exponents—combining together Eqs. (11.11), (11.12) and (11.16)—we obtain the Laplace transform of Eq. (11.2):

$$\mathbf{E}\left[\exp(-\theta \Sigma_1)\right] = \exp\left[-\Gamma(1-\epsilon) c \cdot \theta^\epsilon\right] \,. \tag{11.17}$$

Equation (11.17) implies the following limit:

$$\lim_{\theta \to 0} \frac{1 - \mathbf{E}\left[\exp(-\theta \Sigma_1)\right]}{\theta^\epsilon} = \Gamma(1-\epsilon) c \,. \tag{11.18}$$

Corollary 8.1.7 in [2] asserts that the limit of Eq. (11.18) holds if and only the following limit holds:

$$\lim_{l \to \infty} l^\epsilon \Pr(\Sigma_1 > l) = c \,. \tag{11.19}$$

Equation (11.19) yields the asymptotic equivalence of Eq. (11.3):

$$\Pr(\Sigma_1 > l) \approx c l^{-\epsilon} = \Lambda_-(l) \tag{11.20}$$

(as $l \to \infty$).

11.7.2 Equations (11.5) and (11.6)

Consider the power Poisson process $\tilde{\mathcal{E}}_-$, over the real line, with the intensity function $\tilde{\lambda}_-(y) = \frac{1}{2}\lambda_-(|y|) = \frac{1}{2}c\epsilon |y|^{-\epsilon-1}$ ($-\infty < y < \infty$). Set the test function of

Eq. (11.11) to be $\phi(x) = \exp(i\theta y)$ $(-\infty < y < \infty)$, where θ is a real parameter. Equation (11.4) implies that

$$\mathbf{E}\left[\exp(i\theta \Sigma_2)\right] = \mathbf{E}\left[\exp\left(i\theta \sum_{y \in \tilde{\mathcal{E}}_-} y\right)\right]$$
$$= \mathbf{E}\left[\prod_{y \in \tilde{\mathcal{E}}_-} \exp(i\theta y)\right] = \mathbf{E}\left[\prod_{y \in \tilde{\mathcal{E}}_-} \phi(y)\right]. \tag{11.21}$$

On the other hand, we have

$$\int_{\mathcal{D}} [1 - \phi(y)] \lambda(y) \, dy \tag{11.22}$$

(setting $\mathcal{D} = (-\infty, \infty)$, $\phi(y) = \exp(i\theta y)$, and $\lambda(y) = \tilde{\lambda}_-(y)$)

$$= \int_{-\infty}^{\infty} [1 - \exp(i\theta y)] \frac{1}{2} c\epsilon |y|^{-\epsilon-1} \, dy \tag{11.23}$$

(applying symmetry, integration-by-parts, and a change-of-variables)

$$= 2c |\theta|^\epsilon \int_0^\infty \sin(u) u^{-\epsilon} \, du \, . \tag{11.24}$$

Using a gamma-function calculation, Eqs. (11.22)–(11.24) yield

$$\int_{\mathcal{D}} [1 - \phi(y)] \lambda(y) \, dy = c |\theta|^\epsilon \cdot \frac{\pi}{\Gamma(\epsilon) \sin\left(\frac{1}{2}\pi\epsilon\right)}. \tag{11.25}$$

The integral of Eq. (11.25) is convergent if and only if the exponent ϵ is in the range $0 < \epsilon < 2$. For such exponents—combining together Eqs. (11.11), (11.21) and (11.25)—we obtain the Fourier transform of Eq. (11.5):

$$\mathbf{E}\left[\exp(i\theta \Sigma_2)\right] = \exp\left[-\frac{\pi}{\Gamma(\epsilon) \sin\left(\frac{1}{2}\pi\epsilon\right)} c \cdot |\theta|^\epsilon\right]. \tag{11.26}$$

Equation (11.26) implies the following limit:

$$\lim_{\theta \to 0} \frac{1 - \mathbf{E}\left[\exp(i\theta \Sigma_2)\right]}{|\theta|^\epsilon} = \frac{\pi}{\Gamma(\epsilon) \sin\left(\frac{1}{2}\pi\epsilon\right)} c \, . \tag{11.27}$$

Theorem 8.1.10 in [2] asserts that the limit of Eq. (11.27) holds if and only the following limit holds:

$$\lim_{l \to \infty} l^\epsilon \Pr(\Sigma_2 > l) = c \, . \tag{11.28}$$

Equation (11.28) yields the asymptotic equivalence of Eq. (11.6):

$$\Pr(\Sigma_1 > l) \approx c l^{-\epsilon} = \Lambda_-(l) \tag{11.29}$$

(as $l \to \infty$).

References

1. J.F.C. Kingman, *Poisson Processes* (Oxford University Press, Oxford, 1993)
2. N.H. Bingham, C.M. Goldie, J.L. Teugels, *Regular Variation* (Cambridge University Press, Cambridge, 1987)
3. P. Lévy, Théorie de l'addition des variables Aléatoires (Gauthier-Villars, Paris, 1954)
4. B.V. Gnedenko, A.N. Kolmogorov, *Limit Distributions for Sums of Independent Random Variables* (Addison-Wesley, London, 1954)
5. K.A. Penson, K. Górska, Phys. Rev. Lett. **105**, 210604 (2010)
6. K. Górska, K.A. Penson, Phys. Rev. E **83**, 061125 (2011)
7. S. Redner, *A Guide to First-Passage Processes* (Cambridge University Press, Cambridge, 2001)
8. V. Seshadri, *The Inverse Gaussian Distribution* (Springer, New York, 1998)
9. V.M. Zolotarev, *One-Dimensional Stable Distributions* (American Mathematical Society, Providence, 1986)
10. J.P. Nolan, *Stable Distributions: Models for Heavy Tailed Data* (Birkhauser, New York, 2012)
11. J. Holtsmark, Ann. Phys. **58**, 577 (1919)
12. S. Chandrasekhar, Rev. Mod. Phys. **15**, 1 (1943)

Chapter 12
Dynamics

In this chapter, we explore the nonlinear power dynamics that underpin the power Poisson processes $\mathcal{E}_+$ and $\mathcal{E}_-$. This chapter is based on [1, 2], which study the invariance of Poisson processes with respect to general ODE-based evolution schemes.

12.1 Growth and Decay

Chapter 2 began with the linear ordinary differential equation (ODE) $\dot{X}(t) = r \cdot X(t)$, where $X(t)$ denotes the size (at time t) of some positive quantity of interest, and where r is a real parameter. This linear ODE yields exponential growth for positive r, and yields exponential decay for negative r. We now elevate from the linear ODE to the following nonlinear ODE:

$$\dot{X}(t) = r \cdot \phi[X(t)], \qquad (12.1)$$

where $r \neq 0$ is a real parameter, and where $\phi(x)$ ($x > 0$) is a positive-valued function defined over the positive half-line.

As in the case of the linear ODE, the temporal behavior of the solution of the nonlinear ODE (12.1) is determined by the sign of the parameter r. Indeed, the nonlinear ODE yields a growing solution for positive r, and yields a decaying solution for negative r. In what follows, we denote by $F(x;t)$ the field induced by the nonlinear ODE: $F(x;t)$ is the solution of the nonlinear ODE (12.1) that is generated by the condition $X(0) = x$ ($x > 0$).

As we shall see, the following *power* ODE will turn out to play a key role in this chapter:

$$\frac{\dot{X}(t)}{X(t)} = r \cdot X(t)^p, \qquad (12.2)$$

where p is a real power. The meaning of this power ODE is as follows: the *geometric rate of change* of the quantity of interest—i.e., the quantity's rate of change *relative*

© Springer Nature Switzerland AG 2020
I. Eliazar, *Power Laws*, Understanding Complex Systems,
https://doi.org/10.1007/978-3-030-33235-8_12

to its size—is a power function of its size. This power ODE manifests *allometric growth and decay dynamics* [3].

For $p = 0$, the power ODE (12.2) reduces to the aforementioned linear ODE, whose geometric rate of change is constant: $\dot{X}(t)/X(t) = r$. For $p \neq 0$ the power ODE (12.2) induces the following field:

$$F(x; t) = \left(x^{-p} - rpt\right)^{-1/p}. \tag{12.3}$$

The sign of the term rp—the product of the parameters r and p—determines the temporal range of the field $F(x; t)$ of Eq. (12.3). Specifically: if the product rp is positive then the field's temporal range is $-\infty < t < x^{-p}/rp$; and if the product rp is negative then the field's temporal range is $x^{-p}/rp < t < \infty$.

12.2 Evolution

Consider the following evolution scheme. At time $t = 0$, we initiate from a general Poisson process $\mathcal{P}$ over the positive half-line, with intensity function $\lambda(x)$ ($x > 0$). And, we evolve each and every point of the Poisson process $\mathcal{P}$ according to the dynamics described by the nonlinear ODE (12.1). This evolution scheme combines together stochastic and deterministic features. On the one hand, the initial conditions are stochastic—the points of the Poisson process $\mathcal{P}$. On the other hand, the propagating dynamics are deterministic—driven by the nonlinear ODE (12.1).

Denote by $\mathcal{P}_t$ the collection of points obtained from the evolution scheme at time t ($t > 0$), and recall that $F(x; t)$ denotes the field induced by the nonlinear ODE (12.1). Assume that at time t the term $F(x; -t)$ is positive valued for all $x > 0$. Then, a result from the theory of Poisson processes (the "displacement theorem" [4]) implies that the collection $\mathcal{P}_t$ is a Poisson process over the positive half-line, and that its intensity function is given by

$$\lambda_t(x) = \frac{\lambda[F(x; -t)]\phi[F(x; -t)]}{\phi(x)} \tag{12.4}$$

($x > 0$); the derivation of Eq. (12.4) is detailed in the Methods.

The Poisson process $\mathcal{P}$ is a *fixed point* of the evolution scheme if, for any time $t > 0$, the Poisson process $\mathcal{P}_t$ is equal in law to the initial Poisson process $\mathcal{P}$. Consequently, the Poisson process $\mathcal{P}$ is a fixed point of the evolution scheme if and only if its intensity function is a fixed point of Eq. (12.4): $\lambda_t(x) = \lambda(x)$ ($x > 0$) for all $t > 0$. Substituting $\lambda_t(x) = \lambda(x)$ into Eq. (12.4) implies that the product of the functions $\lambda(x)$ and $\phi(x)$ must be a positive constant:

$$\lambda(x)\phi(x) = const \tag{12.5}$$

($x > 0$).

12.3 Vanishing and Exploding

Equation (12.5) couples together Poisson processes (over the positive half-line) and ODE-based evolution schemes (over the positive half-line). This coupling can be applied in two ways. First, we can set off from the nonlinear ODE (12.1) and obtain the corresponding fixed-point Poisson process $\mathcal{P}$. Second, we can set off from a given Poisson process $\mathcal{P}$ and obtain the corresponding nonlinear ODE, i.e., the ODE for which the Poisson process $\mathcal{P}$ is a fixed point.

Consider the power ODE (12.2), for which $\phi(x) = x^{1+p}$ $(x > 0)$. Equation (12.5) implies that the corresponding fixed-point intensity function is given by: $\lambda(x) = Cx^{-p-1}$ $(x > 0)$, where C is a positive coefficient. This power-law intensity characterizes: the power Poisson process $\mathcal{E}_+$ for negative p; the power Poisson process $\mathcal{E}_-$ for positive p; and the harmonic Poisson process for $p = 0$ [5]. In what follows, we shall focus on the power Poisson processes $\mathcal{E}_+$ and $\mathcal{E}_-$.

12.3 Vanishing and Exploding

Consider the ODE-based evolution scheme of Sect. 12.2 to initiate from the Poisson process $\mathcal{P} = \mathcal{E}_+$. Substituting the intensity function $\lambda(x) = \lambda_+(x) = c\epsilon x^{\epsilon-1}$ $(x > 0)$ into Eq. (12.5) yields the function $\phi(x) = x^{1-\epsilon}$ $(x > 0)$. Consequently, Eq. (12.3) yields the term

$$F(x; -t) = (x^\epsilon - r\epsilon t)^{1/\epsilon}. \tag{12.6}$$

The term appearing in Eq. (12.6) is positive valued, for all $x > 0$ and $t > 0$, if and only if the parameter r is *negative*. With no loss of generality, we set $r = -1/\epsilon$ and obtain a nonlinear ODE that corresponds to the power Poisson process $\mathcal{E}_+$; this ODE, and its induced field, are detailed in Table 12.1. Note that this ODE yields a *decaying* solution that *vanishes* in finite time: if we initiate at time $t = 0$ from the positive value x—then the solution vanishes at the critical time $t_x = x^\epsilon$.

Consider the ODE-based evolution scheme of Sect. 12.2 to initiate from the Poisson process $\mathcal{P} = \mathcal{E}_-$. Substituting the intensity function $\lambda(x) = \lambda_-(x) = c\epsilon x^{-\epsilon-1}$ $(x > 0)$ into Eq. (12.5) yields the function $\phi(x) = x^{1+\epsilon}$ $(x > 0)$. Consequently, Eq. (12.3) yields the term

$$F(x; -t) = \left(x^{-\epsilon} + r\epsilon t\right)^{-1/\epsilon}. \tag{12.7}$$

The term appearing in Eq. (12.7) is positive valued, for all $x > 0$ and $t > 0$, if and only if the parameter r is *positive*. With no loss of generality we set $r = 1/\epsilon$ and obtain a nonlinear ODE that corresponds to the power Poisson process $\mathcal{E}_-$; this ODE, and its induced field, are detailed in Table 12.1. Note that this ODE yields a *growing* solution that *explodes* in finite time: if we initiate at time $t = 0$ from the positive value x—then the solution explodes at the critical time $t_x = x^{-\epsilon}$.

Table 12.1 The ODE-based evolution schemes for which the power Poisson processes, $\mathcal{E}_+$ and $\mathcal{E}_-$, are fixed points. For these Poisson processes, respectively, the rows summarize: the corresponding nonlinear ODEs, their induces fields, and their critical times and critical values

	$\mathcal{E}_+$	$\mathcal{E}_-$
ODE	$\frac{\dot{X}(t)}{X(t)} = \frac{-1}{\epsilon} X(t)^{-\epsilon}$	$\frac{\dot{X}(t)}{X(t)} = \frac{1}{\epsilon} X(t)^{\epsilon}$
Field	$F(x;t) = (x^\epsilon - t)^{1/\epsilon}$	$F(x;t) = (x^{-\epsilon} - t)^{-1/\epsilon}$
Critical time	$t_x = x^\epsilon$	$t_x = x^{-\epsilon}$
Critical value	$F(x;t_x) = 0$	$F(x;t_x) = \infty$

12.4 Order Statistics

As noted in Sect. 12.3, the evolution schemes that correspond to the power Poisson processes, $\mathcal{E}_+$ and $\mathcal{E}_-$, display the following features: the solutions to their corresponding ODEs *vanish* and *explode* in finite time, respectively. Let's now see how these features affect the *order statistics* of the power Poisson processes $\mathcal{E}_+$ and $\mathcal{E}_-$. In what follows, $\{\xi_1, \xi_2, \xi_3, \ldots\}$ are IID copies of an exponential random variable ξ with unit mean.

Consider the order statistics $P_+(1) < P_+(2) < P_+(3) < \cdots$ of the power Poisson process $\mathcal{E}_+$. Evolve each and every order statistic according to the nonlinear ODE that corresponds to the power Poisson process $\mathcal{E}_+$; the details of this ODE are specified in the $\mathcal{E}_+$ column of Table 12.1. The first order statistic $P_+(1)$ will vanish first at the critical time $\tau_+(1) = P_+(1)^\epsilon$. In turn, at the critical time $\tau_+(1)$, the values of all other order statistics shift as follows:

$$P_+(k) \mapsto \tilde{P}_+(k) = F\left[P_+(k); \tau_+(1)\right]$$
$$= \left[P_+(k)^\epsilon - P_+(1)^\epsilon\right]^{1/\epsilon} \tag{12.8}$$

$(k = 2, 3, 4, \ldots)$.

In Chap. 7, we established that the order statistics of the power Poisson process $\mathcal{E}_+$ admit, in law, the representation $P_+(k) = \left[(\xi_1 + \cdots + \xi_k)/c\right]^{1/\epsilon}$ $(k = 1, 2, 3, \ldots)$. Substituting this order-statistics representation into the right-hand side of Eq. (12.8) yields:

$$\tilde{P}_+(k) = \left(\frac{\xi_2 + \cdots + \xi_k}{c}\right)^{1/\epsilon} \tag{12.9}$$

$(k = 2, 3, 4, \ldots)$. Equation (12.9) implies that the sequence $\tilde{P}_+(2) < \tilde{P}_+(3) < \tilde{P}_+(4) < \cdots$ is equal in law to the sequence $P_+(1) < P_+(2) < P_+(3) < \cdots$. Namely, as the first order statistic vanishes ($P_+(1) \mapsto 0$) all the other order statistics ($k = 2, 3, 4, \ldots$) re-form, in law, the original sequence of order statistics.

12.4 Order Statistics

Consider the order statistics $P_-(1) > P_-(2) > P_-(3) > \cdots$ of the power Poisson process $\mathcal{E}_-$. Evolve each and every order statistic according to the nonlinear ODE that corresponds to the power Poisson process $\mathcal{E}_-$; the details of this ODE are specified in the $\mathcal{E}_-$ column of Table 12.1. The first order statistic $P_-(1)$ will explode first at the critical time $\tau_-(1) = P_-(1)^{-\epsilon}$. In turn, at the critical time $\tau_-(1)$, the values of all other order statistics shift as follows:

$$P_-(k) \mapsto \tilde{P}_-(k) = F\left[P_-(k); \tau_-(1)\right]$$
$$= \left[P_-(k)^{-\epsilon} - P_-(1)^{-\epsilon}\right]^{-1/\epsilon} \tag{12.10}$$

($k = 2, 3, 4, \ldots$).

In Chap. 7, we established that the order statistics of the power Poisson process $\mathcal{E}_-$ admit, in law, the representation $P_-(k) = [c/(\xi_1 + \cdots + \xi_k)]^{1/\epsilon}$ ($k = 1, 2, 3, \ldots$). Substituting this order-statistics representation into the right-hand side of Eq. (12.10) yields:

$$\tilde{P}_-(k) = \left(\frac{c}{\xi_2 + \cdots + \xi_k}\right)^{1/\epsilon} \tag{12.11}$$

($k = 2, 3, 4, \ldots$). Equation (12.11) implies that the sequence $\tilde{P}_-(2) > \tilde{P}_-(3) > \tilde{P}_-(4) > \cdots$ is equal in law to the sequence $P_-(1) > P_-(2) > P_-(3) > \cdots$. Namely, as the first order statistic explodes ($P_-(1) \mapsto \infty$) all the other order statistics ($k = 2, 3, 4, \ldots$) re-form, in law, the original sequence of order statistics.

12.5 Beyond the Singularity

In Sect. 12.3, we saw that the evolution schemes that correspond to the power Poisson processes $\mathcal{E}_+$ and $\mathcal{E}_-$ have inherent "singularities"—vanishing and exploding that occur in finite time. In Sect. 12.4, we saw what happens to the power Poisson processes $\mathcal{E}_+$ and $\mathcal{E}_-$ at the singularities that are induced, respectively, by their first order statistics $P_+(1)$ and $P_-(1)$. A natural question that arises is: what happens beyond these "first singularities"?

The answer to this question is: infinitely many singularities, all statistically identical to the "first singularities". Indeed, in Sect. 12.4 we saw that at the first singularities the shifted order statistics re-form, in law, the original sequence of order statistics. Thus, setting off from the first singularities we can generate "second singularities", then "third singularities", and so on and so forth. Moreover, iterating Eqs. (12.9) and (12.11) implies that the order-statistics sequences obtained at these singularities are all identical in law to the original order-statistics sequences.

The "kth singularity" of the power Poisson process $\mathcal{E}_+$ is the epoch at which the kth-order statistic $P_+(k)$ vanishes: the critical time $\tau_+(k) = P_+(k)^{\epsilon}$. Similarly, the "$k$th singularity" of the power Poisson process $\mathcal{E}_-$ is the epoch at which the kth-order statistic $P_-(k)$ explodes: the critical time $\tau_-(k) = P_-(k)^{-\epsilon}$. As noted in Sect. 12.4, and as established in Chap. 7, these order statistics admit the following representations in law: $P_+(k) = [(\xi_1 + \cdots + \xi_k)/c]^{1/\epsilon}$ and $P_-(k) = [c/(\xi_1 + \cdots + \xi_k)]^{1/\epsilon}$, where $\{\xi_1, \xi_2, \xi_3, \ldots\}$ are IID copies of an exponential random variable ξ with unit mean. Thus, we conclude that the singularities of the power Poisson processes $\mathcal{E}_+$ and $\mathcal{E}_-$ occur at the following time epochs:

$$\tau_+(k) = \frac{\xi_1 + \cdots + \xi_k}{c} = \tau_-(k) \qquad (12.12)$$

($k = 1, 2, 3, \ldots$), the equalities being in law.

Equation (12.12) represents, in law, the points of a standard Poisson process with rate c, i.e., a Poisson process, over the positive half-line, with constant intensity c. The points of this process manifest the time epochs of the aforementioned singularities. Thus, we obtain that the waiting times between the consecutive singularities are IID exponential random variables with mean $1/c$.

12.6 Summary

In this chapter, we explored the nonlinear power dynamics that underpin the power Poisson processes $\mathcal{E}_+$ and $\mathcal{E}_-$. At the core of this chapter stand ODE-based decay/growth dynamics. These dynamics give rise to an evolution scheme for Poisson processes over the positive half line: considering a given Poisson process as a collection of initial conditions, and applying the ODE-based dynamics—simultaneously—to all the initial conditions. This evolution scheme combines together stochastic and deterministic features: the initial conditions of the given Poisson process, and the ODE-based dynamics. Also, this evolution scheme gives rise to the notion of fixed points: a Poisson process is a fixed point of the evolution scheme if and only if, statistically, the Poisson process is invariant to the action of the evolution scheme.

Equation (12.5) established a reciprocal connection between decay/growth ODE-based dynamics and Poisson processes over the positive half-line—the reciprocal connection emanating from the fixed points of the aforementioned evolution scheme. In particular, we obtained that: the power Poisson process $\mathcal{E}_+$ corresponds to decaying power dynamics that vanish in finite time; and the power Poisson process $\mathcal{E}_-$ corresponds to growing power dynamics that explode in finite time. Finally, we studied the effect of these vanishing dynamics and explosive dynamics, respectively, on the order statistics of the power Poisson processes $\mathcal{E}_+$ and $\mathcal{E}_-$.

12.7 Methods

Consider a Poisson process $\mathcal{P}$ over a real domain $\mathcal{X}$, with intensity function $\lambda_\mathcal{P}(x)$ ($x \in \mathcal{X}$). Further consider a monotone function $y = f(x)$ that maps the real domain $\mathcal{X}$ to another real domain $\mathcal{Y}$. Applying the map $x \mapsto y = f(x)$ to each and every point of the Poisson process $\mathcal{P}$, we obtain the collection of points

12.7 Methods

$$\mathcal{Q} = \{f(x)\}_{x \in \mathcal{P}} \ . \tag{12.13}$$

The "displacement theorem" of the theory of Poisson processes [4] asserts that $\mathcal{Q}$ is a Poisson process over the real domain $\mathcal{Y}$, with intensity function

$$\lambda_{\mathcal{Q}}(y) = \frac{\lambda_{\mathcal{P}}\left[f^{-1}(y)\right]}{\left|f'\left[f^{-1}(y)\right]\right|} \tag{12.14}$$

($y \in \mathcal{Y}$).

Now, consider the nonlinear ODE (12.1), and set $\Phi(x)$ to be the primitive of the function $1/\phi(x)$ over the positive half-line, i.e., $\Phi'(x) = 1/\phi(x)$ ($x > 0$). As the function $\phi(x)$ is positive valued, the function $\Phi(x)$ is monotone increasing, and so is its inverse function $\Phi^{-1}(y)$. Integrating the ODE (12.1) yields the field

$$F(x; t) = \Phi^{-1}\left[\Phi(x) + rt\right] \ . \tag{12.15}$$

Fix a positive time t, set the domain $\mathcal{X}$ to be the positive half-line, and set $f(x) = F(x; t)$ ($x > 0$). For this setting the domain $\mathcal{Y}$ is the positive half-line, and

$$f^{-1}(y) = \Phi^{-1}\left[\Phi(y) - rt\right] = F(y; -t) \tag{12.16}$$

($y > 0$). In what follows, we implicitly assume that $F(y; -t)$ is positive valued for all $y > 0$. As $\Phi'(x) = 1/\phi(x)$, Eq. (12.16) implies that

$$\frac{1}{f'\left[f^{-1}(y)\right]} = \left[f^{-1}\right]'(y) = \frac{\phi\left[F(y; -t)\right]}{\phi(y)} \tag{12.17}$$

($y > 0$). Substituting Eqs. (12.16) and (12.17) into Eq. (12.14) yields

$$\lambda_{\mathcal{Q}}(y) = \frac{\lambda_{\mathcal{P}}\left[F(y; -t)\right] \phi\left[F(y; -t)\right]}{\phi(y)} \tag{12.18}$$

($y > 0$). Setting $\lambda_{\mathcal{P}}(x) = \lambda(x)$ ($x > 0$), $\mathcal{Q} = \mathcal{P}_t$, and $\lambda_{\mathcal{Q}}(x) = \lambda_t(x)$ ($x > 0$)— Eq. (12.18) yields Eq. (12.4).

References

1. I. Eliazar, J. Klafter, Phys. Rev. E **79**, 011103 (2009)
2. I. Eliazar, J. Phys. A: Math. Theor. **45**, 415103 (2012)
3. L.M. Bettencourt, J. Lobo, D. Helbing, C. Kuhnert, G.B. West, Proc. Natl. Acad. Sci. (USA) **104**, 7301 (2007)
4. J.F.C. Kingman, *Poisson Processes* (Oxford University Press, Oxford, 1993)
5. I. Eliazar, Ann. Phys. **380**, 168 (2017)

Chapter 13
Limit Laws

In Chap. 2, we established a "lognormal" limit law yielding the power Poisson processes $\mathcal{E}_+$ and $\mathcal{E}_-$. In this chapter, we present five more limit laws that yield these power Poisson processes: three limit laws that are based on the linear renormalization schemes of Chap. 10, and two limit laws that are based on the power evolution schemes of Chap. 12.

13.1 Limit Laws I

In Chap. 10, we introduced the linear renormalization schemes that correspond to the power Poisson processes $\mathcal{E}_+$ and $\mathcal{E}_-$. What if we apply these linear renormalization schemes to a general Poisson process $\mathcal{P}$ over the positive half-line? In this section, we address this question. As in Chap. 10, we denote by $\lambda(x)$ ($x > 0$) the intensity function of the general Poisson process $\mathcal{P}$.

Recall that a linear renormalization scheme operates as follows: generate n IID copies of the Poisson process $\mathcal{P}$, take the union of the n IID copies, and then apply the change-of-scale $x \mapsto s \cdot x$ (where s is a positive scale) to each and every point of the union. The result of the linear renormalization scheme is a new Poisson process, over the positive half-line, with intensity function $(n/s) \lambda(x/s)$ ($x > 0$).

As established in Chap. 10, the linear renormalization scheme that corresponds to the power Poisson process $\mathcal{E}_+$ is with scaling $s = n^{1/\epsilon}$. Applying this linear renormalization scheme to the Poisson process $\mathcal{P}$ yields a new Poisson process $\mathcal{P}_n$ with intensity function

$$\lambda_n(x) = n^{1-1/\epsilon} \lambda\left(n^{-1/\epsilon} x\right) \tag{13.1}$$

($x > 0$). Setting $l = n^{-1/\epsilon} x$, Eq. (13.1) implies that a non-trivial limit $\lim_{n \to \infty} \lambda_n(x)$ ($x > 0$) exists if and only if the following limit exists:

$$\lim_{l \to 0} \frac{\lambda(l)}{l^{\epsilon - 1}} = \omega_+ , \tag{13.2}$$

© Springer Nature Switzerland AG 2020
I. Eliazar, *Power Laws*, Understanding Complex Systems,
https://doi.org/10.1007/978-3-030-33235-8_13

where ω_+ is a positive limit value. In turn, if Eq. (13.2) holds then $\lim_{n\to\infty} \lambda_n(x) = \omega_+ x^{\epsilon-1}$ ($x > 0$); this limit is, in effect, the intensity function $\lambda_+(x)$ ($x > 0$) of the power Poisson process $\mathcal{E}_+$. Consequently, the Poisson process $\mathcal{P}_n$ converges in law, in the limit $n \to \infty$, to the power Poisson process $\mathcal{E}_+$.

Hence, we arrive at the two following conclusions. (I) The linear renormalization scheme of Eq. (13.1) leads, in the limit $n \to \infty$, to the power Poisson process $\mathcal{E}_+$. (II) Eq. (13.2) implies that the *domain of attraction* of this linear renormalization scheme comprises all Poisson processes $\mathcal{P}$ whose intensity functions are asymptotically equivalent, near zero, to the intensity function of the power Poisson process $\mathcal{E}_+$: $\lambda(x) \approx \lambda_+(x)$ ($x \to 0$).

As established in Chap. 10, the linear renormalization scheme that corresponds to the power Poisson process $\mathcal{E}_-$ is with scaling $s = n^{-1/\epsilon}$. Applying this linear renormalization scheme to the Poisson process $\mathcal{P}$ yields a new Poisson process $\mathcal{P}_n$ with intensity function

$$\lambda_n(x) = n^{1+1/\epsilon} \lambda\left(n^{1/\epsilon} x\right) \qquad (13.3)$$

($x > 0$). Setting $l = n^{1/\epsilon} x$, Eq. (13.3) implies that a non-trivial limit $\lim_{n\to\infty} \lambda_n(x)$ ($x > 0$) exists if and only if the following limit exists:

$$\lim_{l\to\infty} \frac{\lambda(l)}{l^{-\epsilon-1}} = \omega_-, \qquad (13.4)$$

where ω_- is a positive limit value. In turn, if Eq. (13.4) holds then $\lim_{n\to\infty} \lambda_n(x) = \omega_- x^{-\epsilon-1}$ ($x > 0$); this limit is, in effect, the intensity function $\lambda_-(x)$ ($x > 0$) of the power Poisson process $\mathcal{E}_-$. Consequently, the Poisson process $\mathcal{P}_n$ converges in law, in the limit $n \to \infty$, to the power Poisson process $\mathcal{E}_-$.

Hence, we arrive at the two following conclusions. (I) The linear renormalization scheme of Eq. (13.3) leads, in the limit $n \to \infty$, to the power Poisson process $\mathcal{E}_-$. (II) Eq. (13.4) implies that the *domain of attraction* of this linear renormalization scheme comprises all Poisson processes $\mathcal{P}$ whose intensity functions are asymptotically equivalent, near infinity, to the intensity function of the power Poisson process $\mathcal{E}_-$: $\lambda(x) \approx \lambda_-(x)$ ($x \to \infty$).

13.2 Limit Laws II

In the linear renormalization schemes of Sect. 13.1, we set off from a general Poisson process $\mathcal{P}$ (over the positive half-line), generated n IID copies of it, and took the union of these n IID copies. In this section, we shift from the Poisson-process setting of Sect. 13.1 to an analogous random-variable setting: we replace the general Poisson process $\mathcal{P}$ by a general positive-valued random variable X. Specifically, in this section we set off from the random variable X, generate n IID copies of it, and take the union of these n IID copies: the ensemble $\{X_1, \ldots, X_n\}$. In what follows, $f(x)$ ($x > 0$) denotes the probability density function of the random variable X.

13.2 Limit Laws II

Consider the linear renormalization scheme that corresponds to the power Poisson process $\mathcal{E}_+$; this scheme is with scaling $s = n^{1/\epsilon}$. Applying this scaling to the ensemble $\{X_1, \ldots, X_n\}$ yields the scaled ensemble

$$\mathcal{E}_n = \{n^{1/\epsilon} X_1, \ldots, n^{1/\epsilon} X_n\}. \tag{13.5}$$

The ensemble $\mathcal{E}_n$ converges in law, in the limit $n \to \infty$, to a non-trivial limiting ensemble if and only if the following limit exists:

$$\lim_{l \to 0} \frac{f(l)}{l^{\epsilon-1}} = \omega_+, \tag{13.6}$$

where ω_+ is a positive limit value. In turn, if Eq. (13.6) holds then the limiting ensemble is the power Poisson process $\mathcal{E}_+$. The proof of this limit-law result is detailed in the Methods.

Hence, we arrive at the two following conclusions. (I) The limit-law result regarding the ensemble $\mathcal{E}_n$ of Eq. (13.5) leads, in the limit $n \to \infty$, to the power Poisson process $\mathcal{E}_+$. (II) Eq. (13.6) implies that the *domain of attraction* of this limit-law result comprises all random variables X whose probability density functions $f(x)$ are asymptotically equivalent, near zero, to the intensity function of the power Poisson process $\mathcal{E}_+$: $f(x) \approx \lambda_+(x)$ $(x \to 0)$.

Consider the linear renormalization scheme that corresponds to the power Poisson process $\mathcal{E}_-$; this scheme is with scaling $s = n^{-1/\epsilon}$. Applying this scaling to the ensemble $\{X_1, \ldots, X_n\}$ yields the scaled ensemble

$$\mathcal{E}_n = \{n^{-1/\epsilon} X_1, \ldots, n^{-1/\epsilon} X_n\}. \tag{13.7}$$

The ensemble $\mathcal{E}_n$ converges in law, in the limit $n \to \infty$, to a non-trivial limiting ensemble if and only if the following limit exists:

$$\lim_{l \to \infty} \frac{f(l)}{l^{-\epsilon-1}} = \omega_-, \tag{13.8}$$

where ω_- is a positive limit value. In turn, if Eq. (13.8) holds then the limiting ensemble is the power Poisson process $\mathcal{E}_-$. The proof of this limit-law result is detailed in the Methods.

Hence, we arrive at the two following conclusions. (I) The limit-law result regarding the ensemble $\mathcal{E}_n$ of Eq. (13.7) leads, in the limit $n \to \infty$, to the power Poisson process $\mathcal{E}_-$. (II) Eq. (13.8) implies that the *domain of attraction* of this limit-law result comprises all random variables X whose probability density functions $f(x)$ are asymptotically equivalent, near infinity, to the intensity function of the power Poisson process $\mathcal{E}_-$: $f(x) \approx \lambda_-(x)$ $(x \to \infty)$.

13.3 Limit Laws III

The scaled ensembles of Eqs. (13.5) and (13.7) are special cases of the following scaled ensemble:

$$\mathcal{E}_{n,s} = \{sX_1, \ldots, sX_n\} \, . \tag{13.9}$$

Specifically, in Eq. (13.9): $\{X_1, \ldots, X_n\}$ are n IID copies of a general positive-valued random variable X, and s is an arbitrary positive scale. In particular, setting $s = n^{1/\epsilon}$ in Eq. (13.9) yields Eq. (13.5), and setting $s = n^{-1/\epsilon}$ in Eq. (13.9) yields Eq. (13.7).

In this section, we address the following question: can the number n and the scale s be coupled so that, in the limit $n \to \infty$, we obtain a non-trivial limiting ensemble? Evidently, there are two optional limits for the scale parameter: the limit $s \to \infty$, as in the scaling of Eq. (13.5); and the limit $s \to 0$, as in the scaling of Eq. (13.7).

Consider the case $s \to \infty$, and denote by $F(x) = \Pr(X < x)$ ($x > 0$) the cumulative distribution function of the generic random variable X. The ensemble $\mathcal{E}_{n,s}$ converges in law—in the limits $n \to \infty$ and $s \to \infty$—to a non-trivial limiting ensemble if and only if the two following conditions hold. (I) The number n and the scale s are coupled by the following limit connection:

$$\lim_{n \to \infty, s \to \infty} n F\left(\frac{1}{s}\right) = \omega_+ \, , \tag{13.10}$$

where ω_+ is a positive limit value. (II) The cumulative distribution function $F(x)$ is regularly varying [1] at zero:

$$\lim_{l \to 0} \frac{F(lx)}{F(l)} = x^\epsilon \tag{13.11}$$

($x > 0$), where ϵ is a positive exponent. In turn, if both Eqs. (13.10) and (13.11) hold then the limiting ensemble is the power Poisson process $\mathcal{E}_+$. The proof of this limit-law result is detailed in the Methods.

Hence, we arrive at the two following conclusions. (I) The limit-law result regarding the case $s \to \infty$ yields the power Poisson process $\mathcal{E}_+$. (II) Eq. (13.11) implies that the *domain of attraction* of this limit-law result comprises all random variables X whose cumulative distribution functions are regularly varying at zero. This limit-law result generalizes the corresponding limit-law result of Sect. 13.2, and the domain of attraction of this limit-law result contains the domain of attraction of the corresponding limit-law result: Eq. (13.6) $\Rightarrow$ Eq. (13.11).

Consider the case $s \to 0$, and denote by $\bar{F}(x) = \Pr(X > x)$ ($x > 0$) the tail distribution function of the generic random variable X. The ensemble $\mathcal{E}_{n,s}$ converges in law—in the limits $n \to \infty$ and $s \to 0$—to a non-trivial limiting ensemble if and only if the two following conditions hold. (I) The number n and the scale s are coupled by the following limit connection:

$$\lim_{n \to \infty, s \to 0} n \bar{F}\left(\frac{1}{s}\right) = \omega_- \, , \tag{13.12}$$

13.3 Limit Laws III

where ω_- is a positive limit value. (II) The tail distribution function $\bar{F}(x)$ is regularly varying [1] at infinity:

$$\lim_{l \to \infty} \frac{\bar{F}(lx)}{\bar{F}(l)} = x^{-\epsilon} \tag{13.13}$$

($x > 0$), where ϵ is a positive exponent. In turn, if both Eqs. (13.12) and (13.13) hold then the limiting ensemble is the power Poisson process $\mathcal{E}_-$. The proof of this limit-law result is detailed in the Methods.

Hence, we arrive at the two following conclusions. (I) The limit-law result regarding the case $s \to 0$ yields the power Poisson process $\mathcal{E}_-$. (II) Eq. (13.13) implies that the *domain of attraction* of this limit-law result comprises all random variables X whose tail distribution functions are regularly varying at infinity. This limit-law result generalizes the corresponding limit-law result of Sect. 13.2, and the domain of attraction of this limit-law result contains the domain of attraction of the corresponding limit-law result: Eq. (13.8) $\Rightarrow$ Eq. (13.13).

13.4 Limit Laws IV

In Chap. 12, we pinpointed the ODE-based evolution schemes that correspond to the power Poisson processes $\mathcal{E}_+$ and $\mathcal{E}_-$. What if we apply these evolution schemes to a general Poisson process $\mathcal{P}$ over the positive half-line? In this section, we address this question. As in Chap. 12, we denote by $\lambda(x)$ ($x > 0$) the intensity function of the general Poisson process $\mathcal{P}$.

Recall that an ODE-based evolution scheme operates as follows: initiate, at time $t = 0$, from the points of the Poisson process $\mathcal{P}$; then, propagate each and every point of the Poisson process $\mathcal{P}$ according to the ODE dynamics. The result of the ODE-based evolution scheme, at time t ($t > 0$), is a Poisson process $\mathcal{P}_t$ over the positive half-line. The intensity function of the Poisson process $\mathcal{P}_t$ is:

$$\lambda_t(x) = \lambda[F(x; -t)] \frac{\phi[F(x; -t)]}{\phi(x)} \tag{13.14}$$

($x > 0$), where $\phi(x)$ is the function governing the ODE dynamics, and where $F(x; t)$ is the field induced by the ODE; see Chap. 12 for the details.

As established in Chap. 12, the ODE-based evolution scheme that corresponds to the power Poisson process $\mathcal{E}_+$ is with $\phi(x) = x^{1-\epsilon}$ and with field $F(x; t) = (x^\epsilon - t)^{1/\epsilon}$. In turn, substituting these terms into Eq. (13.14), and setting $l = F(x; -t)$, a straightforward calculation yields

$$\lambda_t(x) = \frac{\lambda(l)}{l^{\epsilon-1}} \cdot x^{\epsilon-1} \tag{13.15}$$

($x > 0$). Noting that $\lim_{t \to \infty} F(x; -t) = \infty$, Eq. (13.15) implies that a non-trivial limit $\lim_{t \to \infty} \lambda_t(x)$ ($x > 0$) exists if and only if the following limit exists:

$$\lim_{l \to \infty} \frac{\lambda(l)}{l^{\epsilon-1}} = \omega_+, \tag{13.16}$$

where ω_+ is a positive limit value. In turn, if Eq. (13.16) holds then $\lim_{t \to \infty} \lambda_t(x) = \omega_+ x^{\epsilon-1}$ ($x > 0$); this limit is, in effect, the intensity function $\lambda_+(x)$ ($x > 0$) of the power Poisson process $\mathcal{E}_+$. Consequently, the Poisson process $\mathcal{P}_t$ converges in law, in the limit $t \to \infty$, to the power Poisson process $\mathcal{E}_+$.

Hence, we arrive at the two following conclusions. (I) The evolution scheme that correspond to the power Poisson process $\mathcal{E}_+$ leads, in the limit $t \to \infty$, to the power Poisson process $\mathcal{E}_+$. (II) Eq. (13.16) implies that the *domain of attraction* of this evolution scheme comprises all Poisson processes $\mathcal{P}$ whose intensity functions are asymptotically equivalent, near infinity, to the intensity function of the power Poisson process $\mathcal{E}_+$: $\lambda(x) \approx \lambda_+(x)$ ($x \to \infty$).

As established in Chap. 12, the ODE-based evolution scheme that corresponds to the power Poisson process $\mathcal{E}_-$ is with $\phi(x) = x^{1+\epsilon}$ and with field $F(x; t) = (x^{-\epsilon} - t)^{-1/\epsilon}$. In turn, substituting these terms into Eq. (13.14), and setting $l = F(x; -t)$, a straightforward calculation yields

$$\lambda_t(x) = \frac{\lambda(l)}{l^{-\epsilon-1}} \cdot x^{-\epsilon-1} \tag{13.17}$$

($x > 0$). Noting that $\lim_{t \to \infty} F(x; -t) = 0$, Eq. (13.17) implies that a non-trivial limit $\lim_{t \to \infty} \lambda_t(x)$ ($x > 0$) exists if and only if the following limit exists:

$$\lim_{l \to 0} \frac{\lambda(l)}{l^{-\epsilon-1}} = \omega_-, \tag{13.18}$$

where ω_- is a positive limit value. In turn, if Eq. (13.18) holds then $\lim_{t \to \infty} \lambda_t(x) = \omega_+ x^{-\epsilon-1}$ ($x > 0$); this limit is, in effect, the intensity function $\lambda_-(x)$ of the power Poisson process $\mathcal{E}_-$ ($x > 0$). Consequently, the Poisson process $\mathcal{P}_t$ converges in law, in the limit $t \to \infty$, to the power Poisson process $\mathcal{E}_-$.

Hence, we arrive at the two following conclusions. (I) The evolution scheme that correspond to the power Poisson process $\mathcal{E}_-$ leads, in the limit $t \to \infty$, to the power Poisson process $\mathcal{E}_-$. (II) Eq. (13.18) implies that the *domain of attraction* of this evolution scheme comprises all Poisson processes $\mathcal{P}$ whose intensity functions are asymptotically equivalent, near zero, to the intensity function of the power Poisson process $\mathcal{E}_-$: $\lambda(x) \approx \lambda_-(x)$ ($x \to 0$).

13.5 Limit Laws V

In Sect. 13.2, we established that the limit-law results of Sect. 13.1—which were devised in the context of Poisson processes—can be modified to the context of random variables. This modification yielded the limit-law results of Sect. 13.2, and thereafter the limit-law results of Sect. 13.3.

It is tempting to modify, in an analogous fashion, the limit-law results of Sect. 13.4. However, an analogous modification will not work. Indeed, the limit-law results of Sect. 13.4 require asymptotic behaviors which *cannot* be met by probability density functions.[1] Nonetheless, the evolution schemes that correspond to the power Poisson processes $\mathcal{E}_+$ and $\mathcal{E}_-$ can be applied in the context of random variables—yet in a setting which is altogether different than that of Sect. 13.4. This different setting, which was introduced and investigated in [2], is presented in this section.

In what follows the ensemble

$$\mathcal{E}_{n,l} = \{lY_1, \ldots, lY_n\} \qquad (13.19)$$

will be the analogue of the ensemble of Eq. (13.9). Specifically, in Eq. (13.19): $\{Y_1, \ldots, Y_n\}$ are n IID copies of a generic positive-valued random variable Y, and l is an arbitrary positive level. Similar to Sect. 13.3, in this section we shall address the following joint limits: $n \to \infty$ and $l \to \infty$; and $n \to \infty$ and $l \to 0$.

Consider the ODE-based evolution scheme that corresponds to the power Poisson process $\mathcal{E}_+$; as noted in Sect. 13.4, this evolution scheme is with field $F(x; t) = (x^\epsilon - t)^{1/\epsilon}$. Also, consider $\{X_1(0), \ldots, X_n(0)\}$ to be n IID copies of a random variable $X(0) = l \cdot X$ where X is a general random variable that takes values in the ray $1 < X < \infty$, with probability density function $f(x)$ ($1 < x < \infty$).

Applying the evolution scheme to the level l yields vanishing at the critical time $t_l = l^\epsilon$. Also, applying the evolution scheme to the initial value $X(0) = l \cdot X$ yields, at the critical time t_l, the random variable $X(t_l) = l \cdot Y$ where

$$Y = (X^\epsilon - 1)^{1/\epsilon}. \qquad (13.20)$$

So, if we start at time $t = 0$ from the ensemble $\{X_1(0), \ldots, X_n(0)\}$, and apply the evolution scheme, then: at the critical time $t = t_l$ we obtain the ensemble $\mathcal{E}_{n,l}$ of Eq. (13.19), with the generic random variable Y given by Eq. (13.20).

The ensemble $\mathcal{E}_{n,l}$ converges in law—in the limits $n \to \infty$ and $l \to \infty$—to a non-trivial limiting ensemble if the two following conditions hold. (I) The number

---

[1] More specifically, the asymptotic equivalences of Sect. 13.4, $\lambda(x) \approx \lambda_+(x)$ ($x \to \infty$) and $\lambda(x) \approx \lambda_-(x)$ ($x \to 0$), imply that the intensity function $\lambda(x)$ ($x > 0$) is non-integrable over the positive half-line: $\int_0^\infty \lambda(x) \, dx = \infty$. Hence, we cannot replace the intensity function $\lambda(x)$ ($x > 0$) with a probability density function $f(x)$ ($x > 0$). The reason why such a replacement worked out in the transition from Sects. 13.1 to 13.2 is that there the asymptotic equivalences were "flipped"—$\lambda(x) \approx \lambda_+(x)$ ($x \to 0$) and $\lambda(x) \approx \lambda_-(x)$ ($x \to \infty$)—thus causing no integrability issue in Sect. 13.2.

n and the level l are coupled by the following limit connection: $\lim_{n\to\infty, l\to\infty} nl^{-\epsilon} = a_+$, where a_+ is a positive limit value. (II) The probability density function $f(x)$ admits a positive left-point limit value: $\lim_{x\to 1} f(x) = b_+$, where b_+ is a positive limit value. In turn, if both these conditions hold then the limiting ensemble is the power Poisson process $\mathcal{E}_+$; the proof of this limit-law result is detailed in the Methods. Note that, as $t_l = l^\epsilon$, the limiting ensemble $\mathcal{E}_+$ is attained at the temporal limit $t_l \to \infty$.

Consider the ODE-based evolution scheme that corresponds to the power Poisson process $\mathcal{E}_-$; as noted in Sect. 13.4, this evolution scheme is with field $F(x; t) = \left(x^{-\epsilon} - t\right)^{-1/\epsilon}$. Also, consider $\{X_1(0), \ldots, X_n(0)\}$ to be n IID copies of a random variable $X(0) = l \cdot X$ where: X is a general random variable that takes values in the unit interval $0 < X < 1$, with probability density function $f(x)$ ($0 < x < 1$).

Applying the evolution scheme to the level l yields explosion at the critical time $t_l = l^{-\epsilon}$. Also, applying the evolution scheme to the initial value $X(0) = l \cdot X$ yields, at the critical time t_l, the random variable $X(t_l) = l \cdot Y$ where

$$Y = \left(X^{-\epsilon} - 1\right)^{-1/\epsilon}. \tag{13.21}$$

So, if we start at time $t = 0$ from the ensemble $\{X_1(0), \ldots, X_n(0)\}$, and apply the evolution scheme, then: at the critical time $t = t_l$ we obtain the ensemble $\mathcal{E}_{n,l}$ of Eq. (13.19), with the generic random variable Y given by Eq. (13.21).

The ensemble $\mathcal{E}_{n,l}$ converges in law—in the limits $n \to \infty$ and $l \to 0$—to a nontrivial limiting ensemble if the two following conditions hold. (I) The number n and the level l are coupled by the following limit connection: $\lim_{n\to\infty, l\to 0} nl^\epsilon = a_-$, where a_- is a positive limit value. (II) The probability density function $f(x)$ admits a positive right-point limit value: $\lim_{x\to 1} f(x) = b_-$, where b_- is a positive limit value. In turn, if both these conditions hold then the limiting ensemble is the power Poisson process $\mathcal{E}_-$; the proof of this limit-law result is detailed in the Methods. Note that, as $t_l = l^{-\epsilon}$, the limiting ensemble $\mathcal{E}_+$ is attained at the temporal limit $t_l \to \infty$.

13.6 Outlook

We conclude with an outlook: an overview of the five limit laws presented along this chapter. The outlook will also encompass the "lognormal" limit law of Chap. 2.

In Chap. 10, we saw that the power Poisson processes $\mathcal{E}_+$ and $\mathcal{E}_-$ are the unique fixed-points of linear renormalization schemes. Armed with this knowledge, in this chapter we explored limit laws emanating from linear renormalization schemes. The first set of limit laws addressed a Poisson-process setting: we established precisely when the application of linear renormalization schemes, on general Poisson processes (over the positive half-line), will yield the power Poisson processes $\mathcal{E}_+$ and $\mathcal{E}_-$. The second set of limit laws addressed a random-variables setting: we modified the linear renormalization schemes to the setting of general positive-valued random variables, and established precisely when these modified renormalizations will yield the power

13.6 Outlook

Poisson processes $\mathcal{E}_+$ and $\mathcal{E}_-$. The third set of limit laws extended the random-variables setting of the second set, and generalized the limit laws of the second set.

In Chap. 12, we saw that the power Poisson processes $\mathcal{E}_+$ and $\mathcal{E}_-$ are the unique fixed-points of power evolution schemes. Armed with this knowledge, in this chapter we also explored limit laws emanating from power evolution schemes. This resulted in two more sets of limit laws. The fourth set of limit laws addressed a Poisson-process setting: we established precisely when the application of power evolution schemes, on general Poisson processes (over the positive half-line), will yield the power Poisson processes $\mathcal{E}_+$ and $\mathcal{E}_-$. The fifth set of limit laws addressed a random-variables setting: we introduced a random-variables setting that is entirely different from the Poisson-process setting of the fourth set, and—for this setting—established precisely when the power evolution schemes will yield the power Poisson processes $\mathcal{E}_+$ and $\mathcal{E}_-$.

So, altogether, we have six limit laws that yield the power Poisson processes $\mathcal{E}_+$ and $\mathcal{E}_-$: the "lognormal" limit law of Chap. 2, and the five limit laws of this chapter. We now turn to discuss and compare these six limit laws.

In Chap. 2 we set off from a positive-valued random variable $X(0)$, and obtained the Poisson-process limits $\mathcal{E}_+$ and $\mathcal{E}_-$ via a *geometric stochastic evolution* with underlying *lognormal statistics*. The domain of attraction of the limit law of Chap. 2 comprises all random variables $X(0)$ with finite moments; this domain of attraction is *wide*. Moreover, in this limit law the "outputs" $\mathcal{E}_+$ and $\mathcal{E}_-$ are determined by the lognormal statistics of the geometric stochastic evolution; we emphasize that the "outputs" are *not* determined by the "input" $X(0)$. Hence, as a mechanism transforming the "input" $X(0)$ to the "outputs" $\mathcal{E}_+$ and $\mathcal{E}_-$, the "lognormal" limit law of Chap. 2 is a *universal mechanism* for the generation of these outputs.

In Sect. 13.2 we set off from a positive-valued random variable X, and obtained the Poisson-process limits $\mathcal{E}_+$ and $\mathcal{E}_-$ via *linear renormalization schemes*. The domains of attraction of the limit laws of Sect. 13.2 comprise all random variables X whose probability density functions follow power asymptotics either near zero or near infinity; these domains of attraction are *narrow*. Moreover, in these limit laws the "outputs" $\mathcal{E}_+$ and $\mathcal{E}_-$ are determined by the power asymptotics of the "input" X. Hence, as a mechanism transforming the "input" X to the "outputs" $\mathcal{E}_+$ and $\mathcal{E}_-$, the limit laws of Sect. 13.2 are "power in, power out" mechanisms: they take the power asymptotics of the "input", and "magnify" these asymptotics to the entire positive half-line. The limit laws of Sect. 13.3 are, in essence, qualitatively identical to the limit laws of Sect. 13.2.

In Sect. 13.5 we set off from a random variable X that is either larger or smaller than the value one, and obtained the Poisson-process limits $\mathcal{E}_+$ and $\mathcal{E}_-$ via *power evolution schemes*. The domains of attraction of the limit laws of Sect. 13.5 comprise all random variables X whose probability density functions admit a positive value at their unit-endpoint; these domains of attraction are *vast*. Moreover, in these limit laws the "outputs" $\mathcal{E}_+$ and $\mathcal{E}_-$ are determined by the power evolution schemes; we emphasize that the "outputs" are *not* determined by the "input" X. Hence, as mechanisms transforming the "input" X into the "outputs" $\mathcal{E}_+$ and $\mathcal{E}_-$, the limit laws of Sect. 13.5 are *universal mechanisms* for the generation of these outputs.

Table 13.1 The domains of attraction of the "lognormal" limit law of Chap. 2, and of the limit laws of Sects. 13.2 and 13.5. In these limit laws, $f(x)$ denotes the probability density function of the underlying "input" random variable. In the case of the "lognormal" limit law, as well as in the case of the limit laws of Sect. 13.2, the density function $f(x)$ is defined over the positive half-line $0 < x < \infty$. In the case of the limit laws of Sect. 13.5, the density function $f(x)$ is defined over: the ray $1 < x < \infty$ for the Poisson-process limit $\mathcal{E}_+$; the unit interval $0 < x < 1$ for the Poisson-process limit $\mathcal{E}_-$. Evidently, the domains of attraction for the three limit laws are determined by markedly different conditions

	$\mathcal{E}_+$	$\mathcal{E}_-$
Chapter 2	$\int_0^\infty x^{-\epsilon} f(x) dx < \infty$	$\int_0^\infty x^\epsilon f(x) dx < \infty$
Section 13.2	$f(x) \underset{x \searrow 0}{\approx} x^{\epsilon-1}$	$f(x) \underset{x \nearrow \infty}{\approx} x^{-\epsilon-1}$
Section 13.5	$f(x) \underset{x \searrow 1}{\approx} 1$	$f(x) \underset{x \nearrow 1}{\approx} 1$

Table 13.2 The domains of attraction of the limit laws of Sects. 13.1 and 13.4. In these limit laws, $\lambda(x)$ denotes the intensity function of the underlying "input" Poisson process, which is defined over the positive half-line $0 < x < \infty$. The intensity functions that belong to the domains of attraction of the limit laws of Sect. 13.1 can be either integrable or not integrable over the positive half-line. The intensity functions that belong to the domains of attraction of the limit laws of Sect. 13.4 are not integrable over the positive half-line

	$\mathcal{E}_+$	$\mathcal{E}_-$
Section 13.1	$\lambda(x) \underset{x \searrow 0}{\approx} x^{\epsilon-1}$	$\lambda(x) \underset{x \nearrow \infty}{\approx} x^{-\epsilon-1}$
Section 13.4	$\lambda(x) \underset{x \nearrow \infty}{\approx} x^{\epsilon-1}$	$\lambda(x) \underset{x \searrow 0}{\approx} x^{-\epsilon-1}$

In Sects. 13.1 and 13.4 we set off from a Poisson process $\mathcal{P}$ over the positive half-line, and: in Sect. 13.1 we obtained the Poisson-process limits $\mathcal{E}_+$ and $\mathcal{E}_-$ via *linear renormalization schemes*; in Sect. 13.4 we obtained the Poisson-process limits $\mathcal{E}_+$ and $\mathcal{E}_-$ via *power evolution schemes*. The domains of attraction of the limit laws of Sect. 13.1, as well as the domains of attraction of the limit laws of Sect. 13.4, comprise all Poisson processes $\mathcal{P}$ whose intensity functions follow power asymptotics either near zero or near infinity; these domains of attraction are *narrow*. Moreover, in these limit laws the "outputs" $\mathcal{E}_+$ and $\mathcal{E}_-$ are determined by the "input" $\mathcal{P}$. Hence, as mechanisms transforming the "input" $\mathcal{P}$ into the "outputs" $\mathcal{E}_+$ and $\mathcal{E}_-$, the limit laws of Sect. 13.1 and the limit laws of Sect. 13.4 are "power in, power out" mechanisms (as described above in the case of the limit laws of Sect. 13.2).

Table 13.1 summarizes the domains of attraction of: the "lognormal" limit law of Chap. 2, the limit laws of Sect. 13.2, and the limit laws of Sect. 13.5. Table 13.2 summarizes the domains of attraction of: the limit laws of Sect. 13.1 and the limit laws of Sect. 13.4.

13.7 Methods

13.7.1 Preparation I

Consider the scaled ensemble

$$\mathcal{E}_{n,s} = \{s \cdot X_1, \ldots, s \cdot X_n\} \tag{13.22}$$

where: s is a positive scale parameter; and $\{X_1, \ldots, X_n\}$ are n IID copies of a general positive-valued random variable X, with probability density function $f(x)$ ($x > 0$). Set $\phi(x)$ ($x > 0$) to be a general, real valued, test function defined over the positive half-line. The characteristic functional of the ensemble $\mathcal{E}_{n,s}$ with respect to the test function $\phi(x)$ is given by

$$\mathbf{E}\left[\prod_{i=1}^{n} \phi(sX_i)\right] = \prod_{i=1}^{n} \mathbf{E}[\phi(sX_i)] = \mathbf{E}[\phi(sX)]^n \ ; \tag{13.23}$$

in Eq. (13.23) we exploited the IID structure of the ensemble $\mathcal{E}_{n,s}$. Note that

$$\mathbf{E}[\phi(sX)] = 1 - \mathbf{E}[1 - \phi(sX)]$$
$$= 1 - \int_0^\infty [1 - \phi(sx)] f(x) dx \tag{13.24}$$

(using the change of variables $x \mapsto y = sx$)

$$= 1 - \int_0^\infty [1 - \phi(y)] \left[\tfrac{1}{s} f\left(\tfrac{y}{s}\right)\right] dy$$
$$= 1 - \tfrac{1}{n} \int_0^\infty [1 - \phi(y)] \left[\tfrac{n}{s} f\left(\tfrac{y}{s}\right)\right] dy \ . \tag{13.25}$$

Hence, setting

$$f_n(y) = \frac{n}{s} f\left(\frac{y}{s}\right) \tag{13.26}$$

($y > 0$), and combining together Eqs. (13.23)–(13.26), we obtain that

$$\mathbf{E}\left[\prod_{i=1}^{n} \phi(sX_i)\right] = \left\{1 - \frac{1}{n} \int_0^\infty [1 - \phi(y)] f_n(y) dy\right\}^n . \tag{13.27}$$

Denote by

$$F(x) = \int_0^x f(y) dy = \Pr(X \leq x) \tag{13.28}$$

($x > 0$) the cumulative distribution function of the random variable X, and note that

$$\int_0^x f_n(y)\,dy = nF\left(\frac{x}{s}\right) \qquad (13.29)$$

($x > 0$). Assuming that $\lim_{x\to\infty}\phi(x) = 1$, Eq. (13.29) implies that

$$\begin{aligned}&\int_0^\infty [1-\phi(y)]\,f_n(y)\,dy \\ &= \int_0^\infty \left[\int_y^\infty \phi'(x)\,dx\right] f_n(y)\,dy \\ &= \int_0^\infty \phi'(x)\left[\int_0^x f_n(y)\,dy\right] dx \\ &= \int_0^\infty \phi'(x)\left[nF\left(\frac{x}{s}\right)\right] dx\,.\end{aligned} \qquad (13.30)$$

Hence, combining together Eqs. (13.27) and (13.30), we obtain that

$$\mathbf{E}\left[\prod_{i=1}^n \phi(sX_i)\right] = \left\{1 - \frac{1}{n}\int_0^\infty \phi'(x)\left[nF\left(\frac{x}{s}\right)\right] dx\right\}^n. \qquad (13.31)$$

Denote by

$$\bar{F}(x) = \int_x^\infty f(y)\,dy = \Pr(X > x) \qquad (13.32)$$

($x > 0$) the tail distribution function of the random variable X, and note that

$$\int_x^\infty f_n(y)\,dy = n\bar{F}\left(\frac{x}{s}\right) \qquad (13.33)$$

($x > 0$). Assuming that $\lim_{x\to 0}\phi(x) = 1$, Eq. (13.33) implies that

$$\begin{aligned}&\int_0^\infty [1-\phi(y)]\,f_n(y)\,dy \\ &= -\int_0^\infty \left[\int_0^y \phi'(x)\,dx\right] f_n(y)\,dy \\ &= -\int_0^\infty \phi'(x)\left[\int_x^\infty f_n(y)\,dy\right] dx \\ &= -\int_0^\infty \phi'(x)\left[n\bar{F}\left(\frac{x}{s}\right)\right] dx\,.\end{aligned} \qquad (13.34)$$

Hence, combining together Eqs. (13.27) and (13.34), we obtain that

$$\mathbf{E}\left[\prod_{i=1}^n \phi(sX_i)\right] = \left\{1 + \frac{1}{n}\int_0^\infty \phi'(x)\left[n\bar{F}\left(\frac{x}{s}\right)\right] dx\right\}^n. \qquad (13.35)$$

13.7 Methods

13.7.2 The Limit $\mathcal{E}_+$ via Eq. (13.5)

The right-hand side of Eq. (13.27) admits a non-trivial limit, as $n \to \infty$, if and only if a non-trivial limit $\lim_{n \to \infty} f_n(y)$ $(y > 0)$ exists. Considering the scaling $s = n^{1/\epsilon}$, Eq. (13.26) implies that

$$\lim_{n \to \infty} f_n(y) = \lim_{n \to \infty} \frac{n}{n^{1/\epsilon}} f\left(\frac{y}{n^{1/\epsilon}}\right) = y^{\epsilon-1} \cdot \lim_{l \to 0} \frac{f(l)}{l^{\epsilon-1}}. \qquad (13.36)$$

Hence, the right-hand side of Eq. (13.27) admits a non-trivial limit, as $n \to \infty$, if and only if a positive limit $\lim_{l \to 0} [l^{1-\epsilon} f(l)]$ exists. Assume that such a positive limit exists: $\lim_{l \to 0} [l^{1-\epsilon} f(l)] = c\epsilon$, where c is a positive constant. Then, Eq. (13.36) implies that

$$\lim_{n \to \infty} f_n(y) = y^{\epsilon-1} \cdot c\epsilon = \lambda_+(y) \qquad (13.37)$$

$(y > 0)$. In turn, Eqs. (13.27) and (13.37) imply that

$$\lim_{n \to \infty} \mathbf{E}\left[\prod_{i=1}^{n} \phi(sX_i)\right] = \exp\left\{-\int_0^\infty [1 - \phi(y)] \lambda_+(y) \, dy\right\}. \qquad (13.38)$$

The right-hand side of Eq. (13.38) is the characteristic functional of a Poisson process, over the positive half-line, with intensity function $\lambda_+(y)$ $(y > 0)$ [3]. Thus, we obtain convergence in law, as $n \to \infty$, to the power Poisson process $\mathcal{E}_+$.

13.7.3 The Limit $\mathcal{E}_-$ via Eq. (13.7)

The right-hand side of Eq. (13.27) admits a non-trivial limit, as $n \to \infty$, if and only if a non-trivial limit $\lim_{n \to \infty} f_n(y)$ $(y > 0)$ exists. Considering the scaling $s = n^{-1/\epsilon}$, Eq. (13.26) implies that

$$\lim_{n \to \infty} f_n(y) = \lim_{n \to \infty} n \cdot n^{1/\epsilon} f\left(n^{1/\epsilon} y\right) = y^{-\epsilon-1} \cdot \lim_{l \to \infty} \frac{f(l)}{l^{-\epsilon-1}}. \qquad (13.39)$$

Hence, the right-hand side of Eq. (13.27) admits a non-trivial limit, as $n \to \infty$, if and only if a positive limit $\lim_{l \to \infty} [l^{1+\epsilon} f(l)]$ exists. Assume that such a positive limit exists: $\lim_{l \to \infty} [l^{1+\epsilon} f(l)] = c\epsilon$, where c is a positive constant. Then, Eq. (13.39) implies that

$$\lim_{n \to \infty} f_n(y) = y^{-\epsilon-1} \cdot c\epsilon = \lambda_-(y) \qquad (13.40)$$

$(y > 0)$. In turn, Eqs. (13.27) and (13.40) imply that

$$\lim_{n\to\infty} \mathbf{E}\left[\prod_{i=1}^{n} \phi(sX_i)\right] = \exp\left\{-\int_0^\infty [1-\phi(y)]\lambda_-(y)\,dy\right\}. \tag{13.41}$$

The right-hand side of Eq. (13.41) is the characteristic functional of a Poisson process, over the positive half-line, with intensity function $\lambda_-(y)$ ($y > 0$) [3]. Thus, we obtain convergence in law, as $n \to \infty$, to the power Poisson process $\mathcal{E}_-$.

13.7.4 The Limit $\mathcal{E}_+$ via Eq. (13.9)

The right-hand side of Eq. (13.31) admits a non-trivial limit, as $n \to \infty$ and $s \to \infty$, if and only if a non-trivial limit $\lim_{n\to\infty, s\to\infty} nF(x/s)$ ($x > 0$) exists. Note that

$$\begin{aligned}
&\lim_{n\to\infty, s\to\infty} nF\left(\frac{x}{s}\right) \\
&= \lim_{n\to\infty, s\to\infty} nF\left(\frac{1}{s}\right) \frac{F(\frac{x}{s})}{F(\frac{1}{s})} \\
&= \lim_{n\to\infty, s\to\infty} nF\left(\frac{1}{s}\right) \cdot \lim_{l\to 0} \frac{F(lx)}{F(l)},
\end{aligned} \tag{13.42}$$

provided that the two limits that appear on the bottom line of Eq. (13.42) exist and are non-trivial.

The first term on the bottom line of Eq. (13.42), $\lim_{n\to\infty, s\to\infty} [nF(1/s)]$, determines the asymptotic coupling between the number n and the scale s:

$$\lim_{n\to\infty, s\to\infty} nF\left(\frac{1}{s}\right) = c, \tag{13.43}$$

where c is a positive constant. The second term on the bottom line of Eq. (13.42), $\lim_{l\to 0}[F(lx)/F(l)]$, characterizes the *regular variation* of the cumulative distribution function $F(x)$ at zero, and it implies that [1]:

$$\lim_{l\to 0} \frac{F(lx)}{F(l)} = x^\epsilon \tag{13.44}$$

($x > 0$), where ϵ is a positive exponent.

If Eqs. (13.43) and (13.44) hold then

$$\lim_{n\to\infty, s\to\infty} nF\left(\frac{x}{s}\right) = cx^\epsilon \tag{13.45}$$

($x > 0$). Combining together Eqs. (13.31) and (13.45) yields

$$\lim_{n\to\infty, s\to\infty} \mathbf{E}\left[\prod_{i=1}^{n} \phi(sX_i)\right] = \exp\left\{-\int_0^\infty \phi'(x)[cx^\epsilon]\,dx\right\}. \tag{13.46}$$

13.7 Methods

Recalling that $\lim_{x \to \infty} \phi(x) = 1$ (in the context of Eq. (13.31)) we have

$$\int_0^\infty \phi'(x) [cx^\epsilon] dx$$
$$= \int_0^\infty \phi'(x) \left[\int_0^x c\epsilon y^{\epsilon-1} dy \right] dx$$
$$= \int_0^\infty \left[\int_y^\infty \phi'(x) dx \right] c\epsilon y^{\epsilon-1} dy \qquad (13.47)$$
$$= \int_0^\infty [1 - \phi(y)] \lambda_+(y) dy .$$

Combining Eqs. (13.46) and (13.47) together we obtain that

$$\lim_{n \to \infty, s \to \infty} \mathbf{E}\left[\prod_{i=1}^n \phi(sX_i) \right] = \exp\left\{ -\int_0^\infty [1 - \phi(y)] \lambda_+(y) dy \right\}. \qquad (13.48)$$

The right-hand side of Eq. (13.48) is the characteristic functional of a Poisson process, over the positive half-line, with intensity function $\lambda_+(y)$ ($y > 0$) [3]. Thus, we obtain convergence in law, as $n \to \infty$ and $s \to \infty$, to the power Poisson process $\mathcal{E}_+$.

13.7.5 The Limit $\mathcal{E}_-$ via Eq. (13.9)

The right-hand side of Eq. (13.35) admits a non-trivial limit, as $n \to \infty$ and $s \to 0$, if and only if a non-trivial limit $\lim_{n \to \infty, s \to 0} n\bar{F}(x/s)$ ($x > 0$) exists. Note that

$$\lim_{n \to \infty, s \to 0} n\bar{F}\left(\frac{x}{s}\right)$$
$$= \lim_{n \to \infty, s \to 0} n\bar{F}\left(\frac{1}{s}\right) \frac{\bar{F}\left(\frac{x}{s}\right)}{\bar{F}\left(\frac{1}{s}\right)} \qquad (13.49)$$
$$= \lim_{n \to \infty, s \to 0} n\bar{F}\left(\frac{1}{s}\right) \cdot \lim_{l \to \infty} \frac{\bar{F}(lx)}{\bar{F}(l)} ,$$

provided that the two limits that appear on the bottom line of Eq. (13.49) exist and are non-trivial. The first term on the bottom line of Eq. (13.49), $\lim_{n \to \infty, s \to 0} \left[n\bar{F}(1/s) \right]$, determines the asymptotic coupling between the number n and the scale s:

$$\lim_{n \to \infty, s \to 0} n\bar{F}\left(\frac{1}{s}\right) = c , \qquad (13.50)$$

where c is a positive constant. The second term on the bottom line of Eq. (13.49), $\lim_{l \to \infty} \left[\bar{F}(lx) / \bar{F}(l) \right]$, characterizes the *regular variation* of the tail distribution function $\bar{F}(x)$ at infinity, and it implies that [1]:

$$\lim_{l \to \infty} \frac{\bar{F}(lx)}{\bar{F}(l)} = x^{-\epsilon} \qquad (13.51)$$

($x > 0$), where ϵ is a positive exponent.

If Eqs. (13.50) and (13.51) hold then

$$\lim_{n \to \infty, s \to 0} n\bar{F}\left(\frac{x}{s}\right) = cx^{-\epsilon} \qquad (13.52)$$

($x > 0$). Combining together Eqs. (13.35) and (13.52) yields

$$\lim_{n \to \infty, s \to 0} \mathbf{E}\left[\prod_{i=1}^{n} \phi(sX_i)\right] = \exp\left\{\int_0^\infty \phi'(x)\left[cx^{-\epsilon}\right]dx\right\}. \qquad (13.53)$$

Recalling that $\lim_{x \to 0} \phi(x) = 1$ (in the context of Eq. (13.35)) we have

$$\begin{aligned}
\int_0^\infty \phi'(x)\left[cx^{-\epsilon}\right]dx \\
= \int_0^\infty \phi'(x)\left[\int_x^\infty c\epsilon y^{-\epsilon-1}dy\right]dx \\
= \int_0^\infty \left[\int_0^y \phi'(x)dx\right]c\epsilon y^{-\epsilon-1}dy \\
= -\int_0^\infty [1 - \phi(y)]\lambda_-(y)\,dy\,.
\end{aligned} \qquad (13.54)$$

Combining Eqs. (13.53) and (13.54) together we obtain that

$$\lim_{n \to \infty, s \to 0} \mathbf{E}\left[\prod_{i=1}^{n} \phi(sX_i)\right] = \exp\left\{-\int_0^\infty [1 - \phi(y)]\lambda_-(y)\,dy\right\}. \qquad (13.55)$$

The right-hand side of Eq. (13.55) is the characteristic functional of a Poisson process, over the positive half-line, with intensity function $\lambda_-(y)$ ($y > 0$) [3]. Thus, we obtain convergence in law, as $n \to \infty$ and $s \to 0$, to the power Poisson process $\mathcal{E}_-$.

13.7.6 Preparation II

Consider the scaled ensemble

$$\mathcal{E}_{n,l} = \{l \cdot Y_1, \ldots, l \cdot Y_n\} \qquad (13.56)$$

where: l is a positive level; and $\{Y_1, \ldots, Y_n\}$ are n IID copies of a general positive-valued random variable Y, with probability density function $g(y)$ ($y > 0$). Set $\phi(y)$ ($y > 0$) to be a general, real valued, test function defined over the positive half-line. Similar to Eqs. (13.26)–(13.27), the characteristic functional of the ensemble $\mathcal{E}_{n,l}$

13.7 Methods

with respect to the test function $\phi(y)$ is given by

$$E\left[\prod_{i=1}^{n}\phi(lY_i)\right] = \left\{1 - \frac{1}{n}\int_0^\infty [1-\phi(y)]g_{n,l}(y)\,dy\right\}^n, \quad (13.57)$$

where

$$g_{n,l}(y) = \frac{n}{l}g\left(\frac{y}{l}\right) \quad (13.58)$$

($y > 0$).

13.7.7 The Limit $\mathcal{E}_+$ via Eq. (13.19)

For the random variable Y of Eq. (13.20) we have

$$\Pr(Y \le y) = \Pr\left[(X^\epsilon - 1)^{1/\epsilon} \le y\right]$$
$$= \Pr\left[X \le (1+y^\epsilon)^{1/\epsilon}\right] \quad (13.59)$$

($y > 0$). Differentiating Eq. (13.59) we obtain the following connection between the density function $g(y)$ of the random variable Y and the density function $f(x)$ of the random variable X:

$$g(y) = f\left[(1+y^\epsilon)^{1/\epsilon}\right](1+y^\epsilon)^{1/\epsilon - 1} y^{\epsilon - 1} \quad (13.60)$$

($y > 0$). Substituting Eq. (13.60) into Eq. (13.58) yields

$$g_{n,l}(y) = \left(nl^{-\epsilon}\right) \cdot f\left\{\left[1+\left(\frac{y}{l}\right)^\epsilon\right]^{1/\epsilon}\right\} \cdot \left[1+\left(\frac{y}{l}\right)^\epsilon\right]^{1/\epsilon - 1} \cdot y^{\epsilon - 1} \quad (13.61)$$

($y > 0$).

Assume that the two following conditions hold. (I) The number n and the level l are coupled by the following limit connection:

$$\lim_{n\to\infty, l\to\infty} nl^{-\epsilon} = a_+, \quad (13.62)$$

where a_+ is a positive limit value. (II) The probability density function $f(x)$ admits a positive left-point limit value:

$$\lim_{x\to 1} f(x) = b_+, \quad (13.63)$$

where b_+ is a positive limit value. Then:

$$\lim_{n\to\infty, l\to\infty} g_{n,l}(y) = a_+ b_+ y^{\epsilon-1} \tag{13.64}$$

($y > 0$). Consequently, Eqs. (13.64) and (13.57) imply that:

$$\lim_{n\to\infty, l\to\infty} E\left[\prod_{i=1}^{n} \phi(lY_i)\right] = \exp\left\{-\int_{0}^{\infty}[1 - \phi(y)]\lambda_+(y)\,dy\right\}, \tag{13.65}$$

where $\lambda_+(y) = c\epsilon y^{\epsilon-1}$ ($y > 0$) and $c = (a_+ b_+)/\epsilon$. The right-hand side of Eq. (13.65) is the characteristic functional of a Poisson process, over the positive half-line, with intensity function $\lambda_+(y)$ ($y > 0$) [3]. Thus, we obtain convergence in law, as $n \to \infty$ and $l \to \infty$, to the power Poisson process $\mathcal{E}_+$.

13.7.8 The Limit $\mathcal{E}_-$ via Eq. (13.19)

For the random variable Y of Eq. (13.21), we have

$$\begin{aligned} \Pr(Y \leq y) &= \Pr\left[\left(X^{-\epsilon} - 1\right)^{-1/\epsilon} \leq y\right] \\ &= \Pr\left[X \leq \left(1 + y^{-\epsilon}\right)^{-1/\epsilon}\right] \end{aligned} \tag{13.66}$$

($y > 0$). Differentiating Eq. (13.66) we obtain the following connection between the density function $g(y)$ of the random variable Y and the density function $f(x)$ of the random variable X:

$$g(y) = f\left[\left(1 + y^{-\epsilon}\right)^{-1/\epsilon}\right]\left(1 + y^{-\epsilon}\right)^{-1/\epsilon - 1} y^{-\epsilon-1} \tag{13.67}$$

($y > 0$). Substituting Eq. (13.67) into Eq. (13.58) yields

$$g_{n,l}(y) = (nl^\epsilon) \cdot f\left\{\left[1 + \left(\frac{l}{y}\right)^\epsilon\right]^{-1/\epsilon}\right\} \cdot \left[1 + \left(\frac{l}{y}\right)^\epsilon\right]^{-1/\epsilon - 1} \cdot y^{-\epsilon-1} \tag{13.68}$$

($y > 0$).

Assume that the two following conditions hold. (I) The number n and the level l are coupled by the following limit connection:

$$\lim_{n\to\infty, l\to 0} nl^\epsilon = a_-, \tag{13.69}$$

where a_- is a positive limit value. (II) The probability density function $f(x)$ admits a positive left-point limit value:

13.7 Methods

$$\lim_{x \to 1} f(x) = b_-, \qquad (13.70)$$

where b_- is a positive limit value. Then

$$\lim_{n \to \infty, l \to 0} g_{n,l}(y) = a_- b_- y^{-\epsilon-1} \qquad (13.71)$$

($y > 0$). Consequently, Eqs. (13.71) and (13.57) imply that

$$\lim_{n \to \infty, l \to 0} \mathbf{E}\left[\prod_{i=1}^{n} \phi(lY_i)\right] = \exp\left\{-\int_0^\infty [1 - \phi(y)] \lambda_-(y) \, dy\right\}, \qquad (13.72)$$

where $\lambda_-(y) = c\epsilon y^{-\epsilon-1}$ ($y > 0$) and $c = (a_- b_-)/\epsilon$. The right-hand side of Eq. (13.72) is the characteristic functional of a Poisson process, over the positive half-line, with intensity function $\lambda_-(y)$ ($y > 0$) [3]. Thus, we obtain convergence in law, as $n \to \infty$ and $l \to 0$, to the power Poisson process $\mathcal{E}_-$.

References

1. N.H. Bingham, C.M. Goldie, J.L. Teugels, *Regular Variation* (Cambridge University Press, Cambridge, 1987)
2. I. Eliazar, Europhys. Lett. **119**, 60007 (2017)
3. J.F.C. Kingman, *Poisson Processes* (Oxford University Press, Oxford, 1993)

Chapter 14
First Digits

In this short chapter, we address the topic of first digits of the power Poisson processes $\mathcal{E}_+$ and $\mathcal{E}_-$. As we shall see, the analysis of the first digits will give rise to power generalizations of the Newcomb–Benford law. This chapter is based on [1], which investigated the first digits and the continued-fractions expansions of Poisson processes over the positive half-line.

The most widely applied method of quantifying the values of positive numbers is the decimal expansion, e.g., Archimedes' constant $\pi = 3.14159\cdots$, Euler's constant $e = 2.71828\cdots$, Planck's constant $h = 6.62607\cdots \times 10^{-34}$, and the gravitational constant $G = 6.67408\cdots \times 10^{-11}$. In the decimal expansion, the first digit of π is 3, the first two digits of e are $(2, 7)$, the first three digits of h are $(6, 6, 2)$, and the first four digits of G are $(6, 6, 7, 4)$. The underlying base of the decimal expansion is 10, and the expansion's digits are $\{0, 1, \ldots, 9\}$.

The decimal expansion is a special case of power expansions. In a general power expansion, the underlying base is an integer b that is greater than one, and the expansion's digits are $\{0, 1, \ldots, b-1\}$. Consider a given vector of m digits, $\mathbf{d} = (d_1, d_2, \ldots, d_m)$, with a non-zero first digit ($d_1 > 0$). Further consider the set of positive numbers whose m first digits—in the base-b power expansion—are the entries of the vector $\mathbf{d}$. This set, which we denote $D(\mathbf{d})$, is given by

$$D(\mathbf{d}) = \bigcup_{i=-\infty}^{\infty} \left\{ \underline{d} \cdot b^i \leq x < \overline{d} \cdot b^i \right\}, \tag{14.1}$$

where $\underline{d} = d_1/b^0 + d_2/b^1 + \cdots + d_m/b^{m-1}$, and where $\overline{d} = \underline{d} + 1/b^{m-1}$.

Our goal in this chapter is to evaluate the "probability" of the set $D(\mathbf{d})$ with respect to the power Poisson processes $\mathcal{E}_+$ and $\mathcal{E}_-$. However, as the power Poisson processes, $\mathcal{E}_+$ and $\mathcal{E}_-$, comprise infinitely many points, we cannot sample at random a point from these processes. In turn, we cannot evaluate the probability that a randomly sampled point takes values in the set $D(\mathbf{d})$. To circumvent this sampling obstacle, we resort to the threshold analysis of Chap. 4.

© Springer Nature Switzerland AG 2020
I. Eliazar, *Power Laws*, Understanding Complex Systems,
https://doi.org/10.1007/978-3-030-33235-8_14

Consider the points of the power Poisson process $\mathcal{E}_+$ that reside below the positive threshold level l. In Chap. 4, we saw that there are finitely many such points, and that these points are IID copies of the inverse-Pareto random variable $B_+(l)$. Set the threshold level l to be a power of the base b, i.e., $l = b^j$, where j is an integer. Then, the probability that the generic random variable $B_+(l)$ takes values in the set $D(\mathbf{d})$ is given by

$$p_+(\mathbf{d}) = \frac{\left(\overline{d}\right)^\epsilon - \left(\underline{d}\right)^\epsilon}{b^\epsilon - 1} ; \qquad (14.2)$$

the derivation of Eq. (14.2) is detailed in the Methods. As the probability appearing in Eq. (14.2) is invariant with respect to the threshold level l, we can take the limit $l \to \infty$ and state that: the fraction of the points of the power Poisson process $\mathcal{E}_+$ whose m first digits are the entries of the vector $\mathbf{d}$ is given by Eq. (14.2).

Consider the points of the power Poisson process $\mathcal{E}_-$ that reside above the positive threshold level l. In Chap. 4, we saw that there are finitely many such points, and that these points are IID copies of the Pareto random variable $A_-(l)$. Set the threshold level l to be a power of the base b, i.e., $l = b^j$, where j is an integer. Then, the probability that the generic random variable $A_-(l)$ takes values in the set $D(\mathbf{d})$ is given by

$$p_-(\mathbf{d}) = \frac{\left(\underline{d}\right)^{-\epsilon} - \left(\overline{d}\right)^{-\epsilon}}{1 - b^{-\epsilon}} ; \qquad (14.3)$$

the derivation of Eq. (14.3) is detailed in the Methods. As the probability appearing in Eq. (14.3) is invariant with respect to the threshold level l, we can take the limit $l \to 0$ and state that: the fraction of the points of the power Poisson process $\mathcal{E}_-$ whose m first digits are the entries of the vector $\mathbf{d}$ is given by Eq. (14.3).

Now, consider the limit $\epsilon \to 0$ in both Eqs. (14.2) and (14.3). L'Hospital's rule implies that the right-hand sides of Eqs. (14.2) and (14.3) yield, in the limit $\epsilon \to 0$, the following probability:

$$p(\mathbf{d}) = \frac{\ln\left(\overline{d}\right) - \ln\left(\underline{d}\right)}{\ln(b)}. \qquad (14.4)$$

Equation (14.4) manifests the fraction of the points of the harmonic Poisson process—which is characterized by the harmonic intensity $\lambda(x) = C/x$ ($x > 0$), where C is a positive coefficient—whose m first digits are the entries of the vector $\mathbf{d}$ [1]. The limit-derivation of the probability of Eq. (14.4) from the probabilities of Eqs. (14.2) and (14.3) is similar, in form, to the limit-derivation of the *Shannon entropy* from the *Rényi entropy* [2–4].

Last, consider the case of the *first digit*, i.e., $m = 1$, and hence $\underline{d} = d_1$ and $\overline{d} = d_1 + 1$. In this case the "probability" that the first digit is *smaller* than the integer i ($i = 1, \ldots, b$) is given by the following formulae: (i)

$$p_+(1) + \cdots + p_+(i-1) = \frac{i^\epsilon - 1}{b^\epsilon - 1} \qquad (14.5)$$

14 First Digits

for the Poisson process $\mathcal{E}_+$; (ii)

$$p_-(1) + \cdots + p_-(i-1) = \frac{1 - i^{-\epsilon}}{1 - b^{-\epsilon}} \qquad (14.6)$$

for the Poisson process $\mathcal{E}_-$; and (iii)

$$p(1) + \cdots + p(i-1) = \frac{\ln(i)}{\ln(b)} \qquad (14.7)$$

for the harmonic Poisson process. Note that setting $i = 1$ in the right-hand sides of Eqs. (14.5)–(14.7) yields a zero value. Also note that setting $\epsilon = 1$ in Eq. (14.5)—the exponent value that characterizes the "standard" Poisson process—yields *uniform* first-digit "probabilities".

Equation (14.7) characterizes the *Newcomb–Benford law* [5, 6], which we discussed in the introduction. Equation (14.4) is the "multi-digit extension" of the Newcomb–Benford law [7, 8]. Equations (14.5) and (14.6) are power generalizations of the Newcomb–Benford law; in effect, the Newcomb–Benford law is to its generalizations what the Shannon entropy is to the Rényi entropy. Equations (14.2) and (14.3) are power generalizations of the "multi-digit extension" of the Newcomb–Benford law.

Methods

Equation (14.2)

Using Eq. (14.1) we have

$$\Pr\left[B_+(l) \in D(\mathbf{d})\right]$$

$$= \Pr\left[\bigcup_{i=-\infty}^{\infty} \{\underline{d} \cdot b^i \leq B_+(l) < \overline{d} \cdot b^i\}\right] \qquad (14.8)$$

$$= \sum_{i=-\infty}^{\infty} \Pr\left[\underline{d} \cdot b^i \leq B_+(l) < \overline{d} \cdot b^i\right]$$

(using the fact that $l = b^j$)

$$= \sum_{i=-\infty}^{j-1} \Pr\left[\underline{d} \cdot b^i \leq B_+(l) < \overline{d} \cdot b^i\right]$$

$$= \sum_{i=-\infty}^{j-1} \Pr\left[\underline{d} \cdot \frac{b^i}{l} \leq \frac{B_+(l)}{l} < \overline{d} \cdot \frac{b^i}{l}\right] \qquad (14.9)$$

$$= \sum_{i=-\infty}^{j-1} \Pr\left[\underline{d} \cdot b^{i-j} \leq \frac{B_+(l)}{l} < \overline{d} \cdot b^{i-j}\right]$$

(applying the change of variables $i \mapsto k = j - i$)

$$= \sum_{k=1}^{\infty} \Pr\left[\underline{d} \cdot b^{-k} \leq \frac{B_+(l)}{l} < \overline{d} \cdot b^{-k}\right]$$
$$= \sum_{k=1}^{\infty} \left\{ \Pr\left[\frac{B_+(l)}{l} < \overline{d} \cdot b^{-k}\right] - \Pr\left[\frac{B_+(l)}{l} < \underline{d} \cdot b^{-k}\right] \right\} .$$
(14.10)

In Chap. 4, we established that

$$\Pr\left[\frac{B_+(l)}{l} < u\right] = u^\epsilon \tag{14.11}$$

$(0 < u < 1)$. Equation (14.11) implies that

$$= \sum_{k=1}^{\infty} \left\{ \Pr\left[\frac{B_+(l)}{l} < \overline{d} \cdot b^{-k}\right] - \Pr\left[\frac{B_+(l)}{l} < \underline{d} \cdot b^{-k}\right] \right\}$$
$$= \sum_{k=1}^{\infty} \left\{ [\overline{d} \cdot b^{-k}]^\epsilon - [\underline{d} \cdot b^{-k}]^\epsilon \right\}$$
$$= \left[(\overline{d})^\epsilon - (\underline{d})^\epsilon\right] \sum_{k=1}^{\infty} (b^{-\epsilon})^k$$
$$= \left[(\overline{d})^\epsilon - (\underline{d})^\epsilon\right] / (b^\epsilon - 1) .$$
(14.12)

Thus, combining Eqs. (14.8)–(14.10) and Eq. (14.12) together we obtain

$$\Pr\left[B_+(l) \in D(\mathbf{d})\right] = \frac{(\overline{d})^\epsilon - (\underline{d})^\epsilon}{b^\epsilon - 1} ; \tag{14.13}$$

this proves Eq. (14.2).

Equation (14.3)

Using Eq. (14.1), we have

$$\Pr\left[A_-(l) \in D(\mathbf{d})\right]$$
$$= \Pr\left[\bigcup_{i=-\infty}^{\infty} \{\underline{d} \cdot b^i \leq A_-(l) < \overline{d} \cdot b^i\}\right] \tag{14.14}$$
$$= \sum_{i=-\infty}^{\infty} \Pr\left[\underline{d} \cdot b^i \leq A_-(l) < \overline{d} \cdot b^i\right]$$

(using the fact that $l = b^j$)

$$= \sum_{i=j}^{\infty} \Pr\left[\underline{d} \cdot b^i \leq A_-(l) < \overline{d} \cdot b^i\right]$$
$$= \sum_{i=j}^{\infty} \Pr\left[\underline{d} \cdot \frac{b^i}{l} \leq \frac{A_-(l)}{l} < \overline{d} \cdot \frac{b^i}{l}\right] \tag{14.15}$$
$$= \sum_{i=j}^{\infty} \Pr\left[\underline{d} \cdot b^{i-j} \leq \frac{A_-(l)}{l} < \overline{d} \cdot b^{i-j}\right]$$

(applying the change of variables $i \mapsto k = i - j$)

$$\begin{aligned}
&= \sum_{k=0}^{\infty} \Pr\left[\underline{d} \cdot b^k \leq \frac{A_-(l)}{l} < \overline{d} \cdot b^k\right] \\
&= \sum_{k=0}^{\infty} \left\{\Pr\left[\frac{A_-(l)}{l} \geq \underline{d} \cdot b^k\right] - \Pr\left[\frac{A_-(l)}{l} \geq \overline{d} \cdot b^k\right]\right\}.
\end{aligned} \quad (14.16)$$

In Chap. 4, we established that

$$\Pr\left[\frac{A_-(l)}{l} > u\right] = u^{-\epsilon} \quad (14.17)$$

$(1 < u < \infty)$. Equation (14.17) implies that

$$\begin{aligned}
&= \sum_{k=0}^{\infty} \left\{\Pr\left[\frac{A_-(l)}{l} \geq \underline{d} \cdot b^k\right] - \Pr\left[\frac{A_-(l)}{l} \geq \overline{d} \cdot b^k\right]\right\} \\
&= \sum_{k=1}^{\infty} \left\{\left[\underline{d} \cdot b^k\right]^{-\epsilon} - \left[\overline{d} \cdot b^k\right]^{-\epsilon}\right\} \\
&= \left[(\underline{d})^{-\epsilon} - (\overline{d})^{-\epsilon}\right] \sum_{k=1}^{\infty} \left(b^{-\epsilon}\right)^k \\
&= \left[(\underline{d})^{-\epsilon} - (\overline{d})^{-\epsilon}\right] / \left(1 - b^{-\epsilon}\right).
\end{aligned} \quad (14.18)$$

Thus, combining Eqs. (14.14)–(14.16) and Eq. (14.18) together we obtain

$$\Pr\left[A_-(l) \in D(\mathbf{d})\right] = \frac{(\underline{d})^{-\epsilon} - (\overline{d})^{-\epsilon}}{1 - b^{-\epsilon}}; \quad (14.19)$$

this proves Eq. (14.3).

References

1. I. Eliazar, Phys. A **392**, 3360 (2013)
2. C.E. Shannon, Bell Syst. Tech. J. **27**, 379 (1948)
3. A. Rényi, *Proceedings of 4th Berkeley Symposium on Mathematical Statistics and Probability*, vol. 1960 (1961), p. 547
4. T.M. Cover, *Elements of Information Theory*, 2nd edn. (Wiley, New York, 2006)
5. A.E. Kossovsky, *Benford's Law* (World Scientific, Singapore, 2014)
6. A. Berger, T.P. Hill, *An Introduction to Benford's Law* (Princeton University Press, Princeton, 2015)
7. T.P. Hill, Amer. Math. Month. **102**, 322 (1995)
8. M. Nigrini, *Benford's Law: Applications for Forensic Accounting, Auditing, and Fraud Detection* (Wiley, Hoboken, 2012)

Chapter 15
Motions

In Chap. 11, we saw how sums induced by the power Poisson process $\mathcal{E}_-$ yield the one-sided and the symmetric Levy laws. In this chapter, we will show how the power Poisson process $\mathcal{E}_-$, when embedded into sums of random motions, gives rise to the following power statistics: sub-diffusion, super-diffusion, flicker noise, and 1/f noise. This chapter is based on [1–4], which investigated the universal generation of "anomalous statistics" via Poisson randomizations of collections of stochastic processes.

15.1 Aggregation

In this chapter, we employ an aggregation scheme that is described as follows. We set off from a general real-valued random motion $X(t)$ (over a time axis that will be specified hereinafter), and generate a countable collection of IID copies $\{X_k(t)\}$ of the random motion. Then, we associate with copy k a time-pace ω_k, where ω_k is a positive parameter, and we run this copy according to the time-pace ω_k; we do so for each and every copy. Finally, summing up all copies we produce the aggregate random motion

$$Y(t) = \sum_k X_k(\omega_k t) \ . \qquad (15.1)$$

Equation (15.1) manifests a mapping whose input is the random motion $X(t)$, whose output is the aggregate random motion $Y(t)$, and whose parameters are the collection of time paces $\mathcal{P} = \{\omega_k\}$. Differentiating Eq. (15.1) with respect to the time variable t yields

$$\dot{Y}(t) = \sum_k \dot{X}_k(\omega_k t) \omega_k \ . \qquad (15.2)$$

Similarly to Eq. (15.1), Eq. (15.2) manifests a mapping whose input is the velocity $\dot{X}(t)$ of the random motion $X(t)$, whose output is the velocity $\dot{Y}(t)$ of the aggregate random motion $Y(t)$, and whose parameters are the collection of time paces $\mathcal{P} = \{\omega_k\}$.

Throughout this chapter, we consider the collection $\mathcal{P}$ to be a Poisson process over the positive half-line, with intensity function $\lambda(\omega)$ ($\omega > 0$). Also, we consider the collection $\mathcal{P}$ to be independent of the IID copies $\{X_k(t)\}$ of the random motion $X(t)$. Consequently, the collection $\mathcal{P}$ is also independent of the IID copies $\{\dot{X}_k(t)\}$ of the velocity $\dot{X}(t)$ of the random motion $X(t)$.

15.2 Diffusive Motions

In this section, we describe diffusions. To that end we set the time axis to be the non-negative half-line ($t \geq 0$), and address real-valued random motions defined over this time axis. For such a random motion, $Z(t)$ ($t \geq 0$), we use the following notation: $m_Z(t) = \mathbf{E}[Z(t)]$ ($t \geq 0$) is the motion's temporal mean function, and $v_Z(t) = \mathbf{Var}[Z(t)]$ ($t \geq 0$) is the motion's temporal variance function.

To describe diffusions, we consider random motions that initiate at the origin and that have a zero mean: $Z(0) = 0$ and $m_Z(t) = 0$ ($t \geq 0$). Consequently, the motion's temporal variance function coincides with its temporal Mean-Square-Displacement (MSD) function:

$$v_Z(t) = \mathbf{E}[|Z(t) - Z(0)|^2] \quad (15.3)$$

($t \geq 0$), where the displacement is measured relative to the motion's initial position ($Z(0) = 0$). Equation (15.3) follows straightforwardly from the fact that the motion's temporal mean function is identically zero.

A diffusion is a random motion that diffuses as time progresses, i.e., whose MSD grows with time. Thus, we say that the random motion $Z(t)$ is a *diffusion* if its temporal variance function $v_Z(t)$ is monotone increasing and divergent, $\lim_{t \to \infty} v_Z(t) = \infty$. In Chap. 2, we encountered the paradigmatic model of diffusion processes in the physical sciences— Brownian motion—whose variance function is linear [5–7]. More generally, a diffusive random motion $Z(t)$ is a *regular diffusion* if its temporal variance function is linear: $v_Z(t) = d_Z \cdot t$ ($t \geq 0$), where d_Z is a positive parameter termed *diffusion coefficient*.

Diffusions with a *power* temporal growth of their MSD are termed *sub-diffusion* and *super-diffusion* [8–10]. Specifically, consider a diffusive random motion $Z(t)$ with a power temporal variance function: $v_Z(t) = c_Z \cdot t^p$ ($t \geq 0$), where c_Z is a positive coefficient, and where $p \neq 1$ is a positive power. Then, sub-linear powers $p < 1$ characterize *sub-diffusion*, and super-linear powers $p > 1$ characterize *super-diffusion*. As noted in the introduction, sub-diffusion and super-diffusion are main pillars of the field of *anomalous diffusion* [8–10].

15.3 Regular Diffusion

With regard to the aggregation scheme of Sect. 15.1, consider the setting of Sect. 15.2: the time axis is the non-negative half-line ($t \geq 0$), and the input $X(t)$ initiates at the origin and has a zero mean—$X(0) = 0$ and $m_X(t) = 0$ ($t \geq 0$). Consequently, also the output $Y(t)$ initiates at the origin: $Y(0) = 0$.

Analysis using conditioning and results from the theory of Poisson processes [11] yields two assertions. (I) As the input $X(t)$, the output $Y(t)$ also has a zero mean: $m_Y(t) = 0$ ($t \geq 0$). (II) The connection between the output's temporal variance function $v_Y(t)$ and the input's temporal variance function $v_X(t)$ is given by

$$v_Y(t) = \int_0^\infty v_X(\omega t) \lambda(\omega) d\omega \tag{15.4}$$

($t \geq 0$). The proof of these two assertions is detailed in the Methods.

Now, consider the input $X(t)$ to be a *regular diffusion*. Substituting the input's linear temporal variance function $v_X(t) = d_X \cdot t$ into Eq. (15.4) straightforwardly implies that the output's temporal variance function is also linear: $v_Y(t) = d_Y \cdot t$, with slope

$$d_Y = \left[\int_0^\infty \lambda(\omega) \omega d\omega \right] \cdot d_X . \tag{15.5}$$

Namely, if the input $X(t)$ is a regular diffusion then so is the output $Y(t)$, and the connection between their diffusion coefficients is given by Eq. (15.5). The aggregation scheme of Sect. 15.1 thus maps regular diffusions to regular diffusions.

Equation (15.5) requires that the intensity function $\lambda(\omega)$ have a finite first moment: $\int_0^\infty \lambda(\omega) \omega d\omega < \infty$. According to a mean result from the theory of Poisson processes [11], this first moment is the mean of the sum of the time-pace parameters, $\mathbf{E}[\sum_k \omega_k]$. Note that the first-moment requirement does *not* allow the collection of time paces $\mathcal{P}$ to be the power Poisson process $\mathcal{E}_-$ (as the intensity function of this power Poisson process has an infinite first moment).

15.4 Anomalous Diffusion

Consider the very same setting as in the previous section: the time axis is the non-negative half-line ($t \geq 0$), and the input $X(t)$ initiates at the origin and has a zero mean. Consequently, as established in the previous section, also the output $Y(t)$ initiates at the origin and has a zero mean. Applying the change-of-variables $\omega \mapsto u = \omega t$ to the right-hand side of Eq. (15.4) yields the following representation of the connection between the output's and input's temporal variance functions:

$$v_Y(t) = \frac{1}{t}\int_0^\infty v_X(u)\,\lambda\left(\frac{u}{t}\right)du \tag{15.6}$$

($t \geq 0$).

With Eq. (15.6) at hand, let us address the following *invariance question*: can we set the collection of time paces $\mathcal{P}$ so that the output's temporal variance function $v_Y(t)$ be invariant, up to a multiplicative constant, with respect to the input $X(t)$? To answer this question, we seek an intensity function $\lambda(\omega)$ for which $v_Y(t) = c_Y \cdot v(t)$ ($t \geq 0$), where c_Y is a positive coefficient, and where $v(t)$ is a function that does not depend on the input $X(t)$. It is clear from Eq. (15.6) that such an invariance holds if and only if the time variable t can be factored out of the term $\lambda(u/t)$, i.e., if and only if the intensity $\lambda(\omega)$ is a power function. Further requiring that the output $Y(t)$ be a diffusion leads to the following conclusion: invariance holds if and only if the collection of time paces $\mathcal{P}$ is the power Poisson process $\mathcal{E}_-$; the proof of this conclusion is detailed in the Methods.

Consider the power Poisson process $\mathcal{E}_-$, with its power intensity function $\lambda_-(\omega) = c\epsilon\omega^{-\epsilon-1}$ ($\omega > 0$). Substituting this power intensity function into Eq. (15.6) yields a power temporal variance function of the output:

$$v_Y(t) = \underbrace{\left[c\epsilon \int_0^\infty \frac{v_X(u)}{u^{\epsilon+1}}du\right]}_{c_Y} \cdot t^\epsilon \tag{15.7}$$

($t \geq 0$); the derivation of Eq. (15.7) is detailed in the Methods. Equation (15.7) requires that the input's temporal variance function $v_X(t)$ satisfy the integrability condition $\int_0^\infty v_X(u)\,u^{-\epsilon-1}du < \infty$.

The power temporal variance function of Eq. (15.7) implies that output $Y(t)$ is a *sub-diffusion* if and only if the exponent ϵ is smaller than one, $\epsilon < 1$. The following pair of criteria assures that the coefficient c_Y is finite throughout the sub-diffusion range $\epsilon < 1$. (I) The input's temporal variance function is asymptotically linear near zero: $\lim_{t \to 0}[v_X(t)/t] = a$, where a is a positive limit value. (II) The input's temporal variance function is bounded: $\lim_{t \to \infty} v_X(t) < \infty$. The informal meaning of these criteria is that the input $X(t)$ kicks off as a regular diffusion, and thereafter it "slows down".

Here is a concrete example of producing sub-diffusion from regular diffusion. Given a regular diffusion $Z(t)$ ($t \geq 0$), kill it at the random time T—where T is a positive random variable that is independent of the regular diffusion. This procedure yields the following input: $X(t) = Z(t)$ for $t < T$, and $X(t) = 0$ for $t \geq T$. And, for this input: if the random variable T has a finite mean then the above pair of sub-diffusion criteria are met, and consequently the output $Y(t)$ is a sub-diffusion. See the Methods for the details of this example.

The power temporal variance function of Eq. (15.7) implies that output $Y(t)$ is a *super-diffusion* if and only if the exponent ϵ is larger than one, $\epsilon > 1$. The following pair of criteria assures that the coefficient c_Y is finite in the entire super-diffusion range $\epsilon > 1$. (I) The input's temporal variance function decays faster than

15.4 Anomalous Diffusion

any power function near zero: $\lim_{t\to 0}[v_X(t)/t^q] = 0$, where this limit holds for any power $q > 2$. (II) The input's temporal variance function is asymptotically linear near infinity: $\lim_{t\to\infty}[v_X(t)/t] = a$, where a is a positive limit value. The informal meaning of these criteria is that the input $X(t)$ kicks off "very slowly", and thereafter it behaves like a regular diffusion.

Here is a concrete example of producing super-diffusion from regular diffusion. Given a regular diffusion $Z(t)$ ($t \geq 0$), start it at the random time T—where T is a positive random variable that is independent of the regular diffusion. This procedure yields the following input: $X(t) = 0$ for $t < T$, and $X(t) = Z(t)$ for $t \geq T$. And, for this input: if the random variable T has finite moments of all negative orders then the above pair of super-diffusion criteria are met, and consequently the output $Y(t)$ is a super-diffusion. See the Methods for the details of this example.

15.5 Stationary Velocities

In this section, we describe stationary velocities. To that end we set the time axis to be the real line ($-\infty < t < \infty$), and address real-valued random velocity processes defined over this time axis. As with the random motions of Sect. 15.2, we consider random velocity processes that have a zero mean.

A stationary velocity process, $\dot{Z}(t)$ ($-\infty < t < \infty$), is characterized by a covariance structure of the following form:

$$\mathbf{Cov}\left[\dot{Z}(t_1), \dot{Z}(t_2)\right] = \rho_{\dot{Z}}(|t_1 - t_2|) \quad (15.8)$$

($-\infty < t_1, t_2 < \infty$), where $\rho_{\dot{Z}}(|\Delta|)$ ($-\infty < \Delta < \infty$) is the velocity's temporal auto-covariance function. Equation (15.8) means that the covariance of the velocity's values at times t_1 and t_2—i.e., the covariance of the random variables $\dot{Z}(t_1)$ and $\dot{Z}(t_2)$—depends only on the time-lag $|t_1 - t_2|$ between the two times. In particular, the velocity's temporal variance function is constant: $v_{\dot{Z}}(t) = \rho_{\dot{Z}}(0)$ ($-\infty < t < \infty$).

Often, the covariance structure of a stationary process is analyzed in Fourier space (rather than in the temporal space). To do so, one shifts from the temporal auto-covariance function of the stationary process under consideration to the Fourier transform of the auto-covariance function— which is termed *spectral density* [12–14]. Namely, the spectral density of the stationary velocity process $\dot{Z}(t)$ is given by

$$\hat{\rho}_{\dot{Z}}(\theta) = \int_{-\infty}^{\infty} \exp(i\theta\Delta)\rho_{\dot{Z}}(|\Delta|)d\Delta \quad (15.9)$$

($-\infty < \theta < \infty$). Of course, one can shift back from the spectral density of Eq. (15.9) to the temporal auto-covariance function of Eq. (15.8) via the Fourier inversion formula [12–14].

Many diffusions have stationary velocity processes. In particular, so does Brownian motion—the paradigmatic model of diffusion processes in the physical sciences [5–7]. Indeed, as we noted in Chap. 2, the velocity of Brownian motion is Gaussian white noise. In general, *white noise* is a zero-mean stationary process, whose covariance structure is characterized by a temporal auto-covariance that is a Dirac "delta function". Consequently, the covariance structure of a white-noise stationary velocity process $\dot{Z}(t)$ is characterized by a *flat* spectral density: $\hat{\rho}_{\dot{Z}}(\theta) = c_{\dot{Z}}$ ($-\infty < \theta < \infty$), where $c_{\dot{Z}}$ is a positive constant.

Stationary processes with a *power* spectral density are termed *flicker noise* [15–17]; in particular, stationary processes with a *harmonic* spectral density are termed *1/f noise*. Namely, the covariance structure of a flicker-noise stationary velocity process $\dot{Z}(t)$ is characterized by the power spectral density $\hat{\rho}_{\dot{Z}}(\theta) = c_{\dot{Z}}/|\theta|^p$ ($-\infty < \theta < \infty$), where $c_{\dot{Z}}$ is a positive coefficient, and where p is a positive power; the power $p = 1$ manifests 1/f noise. As noted in the introduction, the scientific interest in flicker noise and in 1/f noise is vast and transdisciplinary [15–17].

The range-of-dependence of a stationary process is determined by the integrability of its temporal auto-covariance function [18–20]. Equation (15.9) implies that $\hat{\rho}_{\dot{Z}}(0) = 2 \int_0^\infty \rho_{\dot{Z}}(u) \, du$. Hence, the stationary velocity process $\dot{Z}(t)$ has *short-range dependence* if $\hat{\rho}_{\dot{Z}}(0)$ is finite, and has *long-range dependence* if $\hat{\rho}_{\dot{Z}}(0)$ is infinite. In particular, white noise has short-range dependence and flicker noise has long-range dependence.

The finiteness of the variance of a stationary process is determined by the integrability of its spectral density. Indeed, Fourier inversion of Eq. (15.9) implies that $\rho_{\dot{Z}}(0) = \frac{1}{\pi} \int_0^\infty \hat{\rho}_{\dot{Z}}(u) \, du$. Hence, the stationary velocity process $\dot{Z}(t)$ has a *finite variance* if the integral $\int_0^\infty \hat{\rho}_{\dot{Z}}(u) \, du$ is finite, and has an *infinite variance* if the integral $\int_0^\infty \hat{\rho}_{\dot{Z}}(u) \, du$ is infinite. In particular, both white noise and flicker noise have infinite variances.

15.6 White Noise

With regard to the aggregation scheme of Sect. 15.1, consider the setting of Sect. 15.5: the time axis is the real line ($-\infty < t < \infty$), and the input's velocity is a zero-mean stationary process. Analysis using conditioning and results from the theory of Poisson processes [11] asserts that the output's velocity $\dot{Y}(t)$ is a stationary process with the two following properties. (I) As the input's velocity $\dot{X}(t)$, the output's velocity $\dot{Y}(t)$ also has a zero mean. (II) The connection between the spectral density $\hat{\rho}_{\dot{Y}}(\theta)$ of the output's velocity and the spectral density $\hat{\rho}_{\dot{X}}(\theta)$ of the input's velocity is given by

$$\hat{\rho}_{\dot{Y}}(\theta) = \int_0^\infty \hat{\rho}_{\dot{X}}\left(\frac{|\theta|}{\omega}\right) \lambda(\omega) \, \omega \, d\omega \qquad (15.10)$$

($-\infty < \theta < \infty$). The proof of these two properties is detailed in the Methods.

15.6 White Noise

Now, consider the input's velocity $\dot{X}(t)$ to be a *white noise*. Substituting the input's flat spectral density $\hat{\rho}_{\dot{X}}(\theta) = c_{\dot{X}}$ into Eq. (15.10) straightforwardly implies that the output's spectral density is also flat: $\hat{\rho}_{\dot{Y}}(\theta) = c_{\dot{Y}}$, with constant

$$c_{\dot{Y}} = \left[\int_0^\infty \lambda(\omega)\omega d\omega\right] \cdot c_{\dot{X}}. \tag{15.11}$$

Namely, if the input's velocity $\dot{X}(t)$ is a white noise then so is the output's velocity $\dot{Y}(t)$, and the connection between their flat spectral densities is given by Eq. (15.11). The aggregation scheme of Sect. 15.1 thus maps white noises to white noises.

As in the case of Eq. (15.5), Eq. (15.11) requires that the intensity function $\lambda(\omega)$ have a finite first moment: $\int_0^\infty \lambda(\omega)\omega d\omega < \infty$. As noted in Sect. 15.3, this first moment is the mean of the sum of the time-pace parameters, $\mathbf{E}[\sum_k \omega_k]$. Also, as noted in Sect. 15.3, the first-moment requirement does *not* allow the collection of time paces $\mathcal{P}$ to be the power Poisson process $\mathcal{E}_-$ (as the intensity function of this power Poisson process has an infinite first moment).

15.7 Flicker Noise

Consider the very same setting as in the previous section: the time axis is the real line ($-\infty < t < \infty$), and the input's velocity $\dot{X}(t)$ is a zero-mean stationary process. Consequently, as established in the previous section, also the output's velocity $\dot{Y}(t)$ is a zero-mean stationary process. Applying the change-of-variables $\omega \mapsto u = |\theta|/\omega$ to the right-hand side of Eq. (15.10) yields the following representation of the connection between the spectral densities of the output's and input's velocities:

$$\hat{\rho}_{\dot{Y}}(\theta) = |\theta|^2 \int_0^\infty \frac{\hat{\rho}_{\dot{X}}(u)}{u^3} \lambda\left(\frac{|\theta|}{u}\right) du \tag{15.12}$$

($-\infty < \theta < \infty$).

With Eq. (15.12) at hand, let us address the following *invariance question*: can we set the collection $\mathcal{P}$ so that the spectral density of the output's velocity $\hat{\rho}_{\dot{Y}}(\theta)$ be invariant, up to a multiplicative constant, with respect to the input's velocity $\dot{X}(t)$? To answer this question we seek an intensity function $\lambda(\omega)$ for which $\hat{\rho}_{\dot{Y}}(\theta) = c_{\dot{Y}} \cdot \rho(\theta)$ ($-\infty < \theta < \infty$), where $c_{\dot{Y}}$ is a positive constant, and where $\rho(\theta)$ is a function that does not depend on the input's velocity $\dot{X}(t)$. It is clear from Eq. (15.12) that such an invariance holds if and only if the Fourier variable $|\theta|$ can be factored out of the term $\lambda(|\theta|/u)$, i.e., if and only if the intensity $\lambda(\omega)$ is a power function. Further requiring that the spectral density of the output's velocity $\hat{\rho}_{\dot{Y}}(\theta)$ be a decreasing function of the absolute-value variable $|\theta|$ leads to the following conclusion: invariance holds if and only if the collection $\mathcal{P}$ is the power Poisson process $\mathcal{E}_-$ with exponent $\epsilon > 1$; the proof of this conclusion is detailed in the Methods.

Consider the power Poisson process $\mathcal{E}_-$, with its power intensity function $\lambda_-(\omega) = c\epsilon\omega^{-\epsilon-1}$ ($\omega > 0$). Substituting this power intensity function into Eq. (15.12) yields a power spectral density of the output's velocity:

$$\hat{\rho}_{\dot{Y}}(\theta) = \underbrace{\left[c\epsilon \int_0^\infty \frac{\hat{\rho}_{\dot{X}}(u)}{u^{2-\epsilon}} du\right]}_{c_{\dot{Y}}} \cdot \frac{1}{|\theta|^{\epsilon-1}} \tag{15.13}$$

($-\infty < \theta < \infty$); the derivation of Eq. (15.13) is detailed in the Methods. It is evident from Eq. (15.13) that the exponent range $\epsilon > 1$ indeed assures that the spectral density of the output's velocity is a decreasing function of the absolute-value variable $|\theta|$. Also, Eq. (15.13) implies that the output's velocity $\dot{Y}(t)$ is *flicker noise* if and only if the exponent is in the range $\epsilon > 1$, and is *1/f noise* if and only if the exponent is $\epsilon = 2$.

Equation (15.13) requires that the spectral density of the input's velocity satisfy the integrability condition $0 < \int_0^\infty \hat{\rho}_{\dot{X}}(u) u^{\epsilon-2} du < \infty$. In particular, for the case of 1/f noise ($\epsilon = 2$) the integrability condition is $0 < \int_0^\infty \hat{\rho}_{\dot{X}}(u) du < \infty$. In Sect. 15.5, we saw that $\int_0^\infty \hat{\rho}_{\dot{X}}(u) du = \pi \rho_{\dot{X}}(0)$, and hence for the case of 1/f noise ($\epsilon = 2$) we state that: the coefficient $c_{\dot{Y}}$ is positive for all input velocities $\dot{X}(t)$ with a finite variance. In Sect. 15.5, we also saw that $\hat{\rho}_{\dot{X}}(0) = 2 \int_0^\infty \rho_{\dot{X}}(u) du$, and hence we further state that: if the input's velocity $\dot{X}(t)$ has a finite variance, and also has short-range dependence, then the coefficient $c_{\dot{Y}}$ is finite for all exponents in the range $1 < \epsilon \leq 2$.

The quintessential example of a stationary process with finite variance and with short-range dependence is *Ornstein–Uhlenbeck* [21, 22]. The temporal autocovariance function of the Ornstein–Uhlenbeck process is exponential, and its spectral density is Lorentzian: $a/(b + |\theta|^2)$ ($-\infty < \theta < \infty$), where a and b are positive parameters. Consequently, for Ornstein–Uhlenbeck input velocities: the coefficient $c_{\dot{Y}}$ is positive for all exponents in the range $1 < \epsilon < 3$.

15.8 Fusion

So far we addressed diffusions and stationary velocities separately—the former in Sects. 15.2–15.4, and the latter in Sects. 15.5–15.7. In this section, we address these two stochastic processes in a "fused" manner. To that end, we consider a random motion $Z(t) = \int_0^t \dot{Z}(t') dt'$ ($t \geq 0$), which is generated via a zero-mean stationary velocity process $\dot{Z}(t)$ ($-\infty < t < \infty$).

In terms of the spectral density $\hat{\rho}_{\dot{Z}}(\theta)$ ($-\infty < \theta < \infty$) of the velocity process $\dot{Z}(t)$, the temporal variance function of the random motion $Z(t)$ is given by

$$v_Z(t) = \frac{2}{\pi} t \int_0^\infty \hat{\rho}_{\dot{Z}}\left(\frac{u}{t}\right) \frac{1 - \cos(u)}{u^2} du \tag{15.14}$$

15.8 Fusion

($t \geq 0$); the derivation of Eq. (15.14) is detailed in the Methods. As we shall now show, Eq. (15.14) yields a path from white-noise velocities to regular-diffusion motions, and a path from flicker-noise velocities to super-diffusive motions.

First, let's apply Eq. (15.14) to the output velocity $\dot{Y}(t)$ that was produced in Sect. 15.6 via the aggregation scheme (with a collection of time paces $\mathcal{P}$ whose intensity function $\lambda(\omega)$ has a finite first moment, $\int_0^\infty \lambda(\omega)\omega d\omega < \infty$). The *flat* spectral density of this output's velocity is given by Eq. (15.11), and substituting this flat spectral density into Eq. (15.14) yields the following *linear* temporal variance function of the random motion $Y(t)$:

$$v_Y(t) = c_{\dot{Y}} \cdot t \tag{15.15}$$

($t \geq 0$), where $c_{\dot{Y}}$ is the coefficient of Eq. (15.11).[1] Namely, Eq. (15.15) implies that the random motion $Y(t)$ is a regular diffusion with diffusion coefficient $d_Y = c_{\dot{Y}}$ (where $c_{\dot{Y}}$ is given by Eq. (15.11)). Thus, as stated above: white-noise velocities lead to regular-diffusion motions.

Second, let's apply Eq. (15.14) to the output velocity $\dot{Y}(t)$ that was produced in Sect. 15.7 via the aggregation scheme (with a collection of time paces $\mathcal{P}$ that is the power Poisson process $\mathcal{E}_-$). The *power* spectral density of this output's velocity is given by Eq. (15.13), and substituting this power spectral density into Eq. (15.14) yields the following *power* temporal variance function of the random motion $Y(t)$:

$$v_Y(t) = \underbrace{\left[\frac{2}{\pi} c_{\dot{Y}} \int_0^\infty \frac{1 - \cos(u)}{u^{\epsilon+1}} du\right]}_{c_Y} \cdot t^\epsilon \tag{15.16}$$

($t \geq 0$), where $c_{\dot{Y}}$ is the coefficient of Eq. (15.13).

The integral $\int_0^\infty [1 - \cos(u)] u^{-\epsilon-1} du$ appearing on the right-hand side of Eq. (15.16) is integrable for all exponents $1 \leq \epsilon < 2$. Consequently, whenever the coefficient $c_{\dot{Y}}$ of Eq. (15.13) is positive for exponents $1 < \epsilon < 2$ we conclude that: the output's velocity $\dot{Y}(t)$ of Sect. 15.7 is a flicker noise, and the output $Y(t)$ is a super-diffusion. For example, this conclusion holds valid for Ornstein–Uhlenbeck input velocities (which were discussed in Sect. 15.7). Thus, as stated above: flicker-noise velocities lead to super-diffusive motions.

15.9 Outlook

In this chapter, we employed an aggregation scheme that can be perceived as a signal-superposition model that is described as follows. There is a countable collection of transmission sources that operate simultaneously and independently of each other.

---

[1] In the transition from Eqs. (15.14) to (15.15) we used the following integral identity: $\int_0^\infty [1 - \cos(u)] u^{-2} du = \pi/2$.

Given a temporal random signal process—the input $X(t)$—the sources transmit IID copies of this signal; however, each source transmits its copy according to its own inherent time pace. Collectively, the sources produce the output $Y(t)$—the aggregate of all their transmissions. Thus, the sources convert a given temporal input random signal $X(t)$ to a temporal output random signal $Y(t)$. And, in this conversion, the sources use a collection of parameters $\mathcal{P}$—their time paces; these time paces are considered independent of the transmitted signals (the input's IID copies).

We set the collection of parameters $\mathcal{P}$ to be a Poisson process over the positive half-line. Doing so, and considering zero-mean inputs that initiate at the origin, we investigated the outputs' temporal variance functions; in particular, we showed that regular-diffusion inputs result in regular-diffusion outputs. Also, considering inputs with zero-mean stationary velocities, we established that so are the outputs' velocities, and we investigated the outputs' spectral densities; in particular, we showed that inputs with white-noise velocities result in outputs with white-noise velocities.

Most importantly, in this chapter we addressed two key *invariance questions*. The first invariance question regarded the setting of zero-mean inputs that initiate at the origin, and asked: can we set the collection of parameters $\mathcal{P}$ so that the outputs' temporal variance functions be invariant, up to a multiplicative factor, of the inputs? The second invariance question regarded the setting of inputs with zero-mean stationary velocities, and asked: can we set the collection of parameters $\mathcal{P}$ so that the spectral densities of the outputs' velocities be invariant, up to a multiplicative factor, of the inputs?

Interestingly, the power Poisson process $\mathcal{E}_-$ arises from both these invariance questions. Indeed, we established that the answer to the first invariance question is affirmative if and only if $\mathcal{P} = \mathcal{E}_-$—in which case the resulting outputs are sub-diffusions and super-diffusions. Also, we established that the answer to the second invariance question is affirmative if and only if $\mathcal{P} = \mathcal{E}_-$—in which case the velocities of the resulting outputs are flicker noises (which, in particular, include the case of 1/f noise).

The first invariance question led to Eq. (15.7). Based on the power Poisson process $\mathcal{E}_-$, Eq. (15.7) is a universal design mechanism for sub-diffusion and super-diffusion. By tuning the parameters of the power Poisson process $\mathcal{E}_-$ (i.e., the coefficient c and the exponent ϵ of its power intensity function), one can produce any desired power temporal variance function of the output $Y(t)$. Once a desired exponent ϵ is set, the admissible inputs $X(t)$ are characterized by a simple integrability condition: the integral appearing on the right-hand side of Eq. (15.7) should be finite.

The second invariance question led to Eq. (15.13). Based on the power Poisson process $\mathcal{E}_-$, Eq. (15.13) is a universal design mechanism for flicker noise. By tuning the parameters of the power Poisson process $\mathcal{E}_-$ (i.e., the coefficient c and the exponent ϵ of its power intensity function), one can produce any desired power spectral density of the output's velocity $\dot{Y}(t)$. Once a desired exponent ϵ is set, the admissible input velocities $\dot{X}(t)$ are characterized by a simple integrability condition: the integral appearing on the right-hand side of Eq. (15.13) should be positive.

15.10 Methods

15.10.1 Equation (15.4)

The conditional mean of the output $Y(t)$, given the information $\mathcal{P} = \{\omega_k\}$, is

$$\mathbf{E}[Y(t) \mid \mathcal{P}] = \mathbf{E}\left[\sum_k X_k(\omega_k t) \mid \mathcal{P}\right]$$

$$= \sum_k \mathbf{E}[X_k(\omega_k t) \mid \mathcal{P}] = \sum_k \mathbf{E}[X(\omega_k t) \mid \mathcal{P}] \quad (15.17)$$

$$= \sum_k m_X(\omega_k t) = 0$$

($t \geq 0$); in Eq. (15.17) we used the setting of the aggregation scheme (described in Sect. 15.1), and the fact that the input's temporal mean function $m_X(t)$ is identically zero.

The conditional variance of the output $Y(t)$, given the information $\mathcal{P} = \{\omega_k\}$, is

$$\mathbf{Var}[Y(t) \mid \mathcal{P}] = \mathbf{Var}\left[\sum_k X_k(\omega_k t) \mid \mathcal{P}\right]$$

$$= \sum_k \mathbf{Var}[X_k(\omega_k t) \mid \mathcal{P}] = \sum_k \mathbf{Var}[X(\omega_k t) \mid \mathcal{P}] \quad (15.18)$$

$$= \sum_k v_X(\omega_k t)$$

($t \geq 0$); in Eq. (15.18) we used the setting of the aggregation scheme (described in Sect. 15.1).

Using conditioning and Eq. (15.17), the output's temporal mean function is given by

$$m_Y(t) = \mathbf{E}[Y(t)]$$

$$= \mathbf{E}[\mathbf{E}[Y(t) \mid \mathcal{P}]] \quad (15.19)$$

$$= \mathbf{E}[0] = 0$$

($t \geq 0$).

Using conditioning and Eqs. (15.17)–(15.18), the output's temporal variance function is given by

$$v_Y(t) = \mathbf{Var}[Y(t)]$$

$$= \mathbf{E}[\mathbf{Var}[Y(t) \mid \mathcal{P}]] + \mathbf{Var}[\mathbf{E}[Y(t) \mid \mathcal{P}]]$$

$$= \mathbf{E}\left[\sum_k v_X(\omega_k t)\right] + \mathbf{Var}[0] \qquad (15.20)$$

$$= \int_0^\infty v_X(\omega t) \lambda(\omega) d\omega$$

($t \geq 0$); in the transition from the third line of Eq. (15.20) to its last line we used a mean result from the theory of Poisson processes—Eq. (3.9) in [11]. This proves Eq. (15.4).

15.10.2 Invariance and Eq. (15.7)

Consider Eq. (15.6) and the invariance question following it. As explained in Sect. 15.4, invariance holds if and only if the intensity function of the Poisson process $\mathcal{P}$ is a power function, i.e.: $\lambda(\omega) = C\omega^{p-1}$ ($\omega > 0$), where C is a positive coefficient, and where p is a real power. Substituting this power intensity function into Eq. (15.6) yields

$$v_Y(t) = \frac{1}{t} \int_0^\infty v_X(u) \lambda\left(\frac{u}{t}\right) du$$

$$= \frac{1}{t} \int_0^\infty v_X(u) \left[C\left(\frac{u}{t}\right)^{p-1}\right] du \qquad (15.21)$$

$$= \underbrace{\left[C \int_0^\infty v_X(u) u^{p-1} du\right]}_{c_Y} \cdot t^{-p}$$

($t \geq 0$). Evidently—provided that c_Y is positive—the temporal variance function of Eq. (15.21) is monotone increasing if and only if the power p is negative. Setting $p = -\epsilon$ and $C = c\epsilon$ in Eq. (15.21) yields Eq. (15.7).

15.10.3 Anomalous-Diffusion Examples

Consider a regular diffusion $Z(t)$ ($t \geq 0$), and a random time T (a positive random variable) that is independent of the regular diffusion. Denote by $f(t)$ ($t > 0$) the density function of the random variable T; and denote by $F(t) = \Pr(T < t)$ ($t \geq 0$) and by $\bar{F}(t) = \Pr(T > t)$ ($t \geq 0$) the corresponding distribution functions. In what follows $I\{E\}$ is the indicator function of an event E, i.e., $I\{E\} = 1$ if the event occurred, and $I\{E\} = 0$ if the event did not occur.

15.10 Methods

Introduce the random motion

$$X(t) = Z(t) \cdot I\{t < T\} \tag{15.22}$$

($t \geq 0$). The properties of the regular diffusion $Z(t)$ and of the random time T imply that

$$\begin{aligned}\mathbf{E}[X(t)] &= \mathbf{E}[Z(t) \cdot I\{t < T\}] \\ &= \mathbf{E}[Z(t)] \cdot \mathbf{E}[I\{t < T\}] \\ &= m_Z(t) \cdot \Pr(t < T) \\ &= 0 \cdot \bar{F}(t) = 0 \, ; \end{aligned} \tag{15.23}$$

and

$$\begin{aligned}\mathbf{E}\left[X(t)^2\right] &= \mathbf{E}\left[Z(t)^2 \cdot I\{t < T\}^2\right] \\ &= \mathbf{E}\left[Z(t)^2\right] \cdot \mathbf{E}[I\{t < T\}] \\ &= v_Z(t) \cdot \Pr(t < T) = (d_Z \cdot t) \cdot \bar{F}(t) \, . \end{aligned} \tag{15.24}$$

Equation (15.23) yields the temporal mean function $m_X(t) = 0$ ($t \geq 0$), and Eqs. (15.23)–(15.24) yield the temporal variance function

$$v_X(t) = d_Z \cdot t \bar{F}(t) \tag{15.25}$$

($t \geq 0$). In turn, for exponents $0 < \epsilon < 1$, Eq. (15.25) implies that

$$\begin{aligned}\int_0^\infty \tfrac{v_X(u)}{u^{\epsilon+1}} du &= d_Z \cdot \int_0^\infty u^{-\epsilon} \bar{F}(u)\, du \\ &= d_Z \cdot \int_0^\infty \tfrac{t^{1-\epsilon}}{1-\epsilon} f(t)\, dt = \tfrac{d_Z}{1-\epsilon} \mathbf{E}\left[T^{1-\epsilon}\right] \end{aligned} \tag{15.26}$$

in the transition from the first line of Eq. (15.26) to its second line we used integration-by-parts. If the random time T has a finite mean then the integral of Eq. (15.26) is finite for all exponents $0 < \epsilon < 1$.

Introduce the random motion

$$X(t) = Z(t) \cdot I\{t \geq T\} \tag{15.27}$$

($t \geq 0$). Similarly to Eqs. (15.23)–(15.24), the temporal mean function is $m_X(t) = 0$ ($t \geq 0$), and the temporal variance function is

$$v_X(t) = d_Z \cdot t F(t) \tag{15.28}$$

($t \geq 0$). In turn, for exponents $\epsilon > 1$, Eq. (15.28) implies that

$$\int_0^\infty \frac{v_X(u)}{u^{\epsilon+1}} du = d_Z \cdot \int_0^\infty u^{-\epsilon} F(u) \, du$$
$$= d_Z \cdot \int_0^\infty \frac{t^{1-\epsilon}}{\epsilon-1} f(t) \, dt = \frac{d_Z}{\epsilon-1} \mathbf{E}\left[T^{1-\epsilon}\right] \tag{15.29}$$

in the transition from the first line of Eq. (15.29) to its second line we used integration-by-parts. If the random variable T has a finite moments of all negative orders then the integral of Eq. (15.29) is finite for all exponents $\epsilon > 1$.

15.10.4 Equation (15.10)

The conditional mean of the output's velocity $\dot{Y}(t)$, given the information $\mathcal{P} = \{\omega_k\}$, is

$$\mathbf{E}\left[\dot{Y}(t) \mid \mathcal{P}\right] = \mathbf{E}\left[\sum_k \dot{X}_k(\omega_k t) \omega_k \mid \mathcal{P}\right]$$
$$= \sum_k \omega_k \mathbf{E}\left[\dot{X}_k(\omega_k t) \mid \mathcal{P}\right] = \sum_k \omega_k \mathbf{E}\left[\dot{X}(\omega_k t) \mid \mathcal{P}\right] \tag{15.30}$$
$$= \sum_k \omega_k m_{\dot{X}}(\omega_k t) = 0$$

$(-\infty < t < \infty)$; in Eq. (15.30) we used the setting of the aggregation scheme (described in Sect. 15.1), and the fact that the temporal mean function $m_{\dot{X}}(t)$ of the input's velocity is identically zero.

The conditional covariance of the output's velocity $\dot{Y}(t)$, given the information $\mathcal{P} = \{\omega_k\}$, is

$$\mathbf{Cov}\left[\dot{Y}(t_1), \dot{Y}(t_2) \mid \mathcal{P}\right]$$
$$= \mathbf{Cov}\left[\sum_k \dot{X}_k(\omega_k t_1) \omega_k, \sum_j \dot{X}_j(\omega_j t_2) \omega_j \mid \mathcal{P}\right]$$
$$= \sum_k \sum_j \omega_k \omega_j \mathbf{Cov}\left[\dot{X}_k(\omega_k t_1), \dot{X}_j(\omega_j t_2) \mid \mathcal{P}\right] \tag{15.31}$$
$$= \sum_k \omega_k^2 \mathbf{Cov}\left[\dot{X}_k(\omega_k t_1), \dot{X}_k(\omega_k t_2) \mid \mathcal{P}\right]$$
$$= \sum_k \omega_k^2 \mathbf{Cov}\left[\dot{X}(\omega_k t_1), \dot{X}(\omega_k t_2) \mid \mathcal{P}\right]$$
$$= \sum_k \omega_k^2 \rho_{\dot{X}}(\omega_k |t_1 - t_2|)$$

$(-\infty < t < \infty)$; in Eq. (15.31) we used the setting of the aggregation scheme (described in Sect. 15.1).

Using conditioning and Eq. (15.30), the temporal mean function of the output's velocity $\dot{Y}(t)$ is given by

15.10 Methods

$$m_{\dot{Y}}(t) = \mathbf{E}\left[\dot{Y}(t)\right]$$
$$= \mathbf{E}\left[\mathbf{E}\left[\dot{Y}(t) \mid \mathcal{P}\right]\right] \quad (15.32)$$
$$= \mathbf{E}[0] = 0$$

$(-\infty < t < \infty)$.

Using conditioning and Eqs. (15.30)–(15.31), the covariance of the output's velocity $\dot{Y}(t)$ is given by

$$\mathbf{Cov}\left[\dot{Y}(t_1), \dot{Y}(t_2)\right]$$
$$= \mathbf{E}\left[\mathbf{Cov}\left[\dot{Y}(t_1), \dot{Y}(t_2) \mid \mathcal{P}\right]\right] + \mathbf{Cov}\left[\mathbf{E}\left[\dot{Y}(t_1) \mid \mathcal{P}\right], \mathbf{E}\left[\dot{Y}(t_2) \mid \mathcal{P}\right]\right]$$
$$= \mathbf{E}\left[\sum_k \omega_k^2 \rho_{\dot{X}}\left(\omega_k \left|t_1 - t_2\right|\right)\right] + \mathbf{Cov}[0, 0] \quad (15.33)$$
$$= \int_0^\infty \omega^2 \rho_{\dot{X}}\left(\omega \left|t_1 - t_2\right|\right) \lambda(\omega)\, d\omega$$

$(-\infty < t < \infty)$; in the transition from the third line of Eq. (15.33) to its last line, we used a mean result from the theory of Poisson processes—Eq. (3.9) in [11].

Equation (15.33) implies that the temporal auto-covariance function of the output's velocity $\dot{Y}(t)$ is

$$\rho_{\dot{Y}}(|\Delta|) = \int_0^\infty \rho_{\dot{X}}(\omega|\Delta|)\,\lambda(\omega)\,\omega^2\,d\omega \quad (15.34)$$

$(-\infty < \Delta < \infty)$. Applying a Fourier transform to Eq. (15.34) yields

$$\hat{\rho}_{\dot{Y}}(\theta) = \int_{-\infty}^\infty \exp(i\theta\Delta)\,\rho_{\dot{Y}}(|\Delta|)\,d\Delta$$
$$= \int_{-\infty}^\infty \exp(i\theta\Delta) \left[\int_0^\infty \rho_{\dot{X}}(\omega|\Delta|)\,\lambda(\omega)\,\omega^2\,d\omega\right] d\Delta \quad (15.35)$$
$$= \int_0^\infty \left[\int_{-\infty}^\infty \exp(i\theta\Delta)\,\rho_{\dot{X}}(|\omega\Delta|)\,d\Delta\right] \lambda(\omega)\,\omega^2\,d\omega$$

$(-\infty < \theta < \infty)$. Using the change-of-variables $\Delta \mapsto \Delta' = \omega\Delta$, note that

$$\int_{-\infty}^\infty \exp(i\theta\Delta)\,\rho_{\dot{X}}(|\omega\Delta|)\,d\Delta$$
$$= \frac{1}{\omega} \int_{-\infty}^\infty \exp\left(i\frac{\theta}{\omega}\Delta'\right) \rho_{\dot{X}}(|\Delta'|)\,d\Delta' \quad (15.36)$$
$$= \frac{1}{\omega}\hat{\rho}_{\dot{X}}\left(\frac{\theta}{\omega}\right)$$

($-\infty < \theta < \infty$). Substituting Eq. (15.36) into Eq. (15.35) yields

$$\hat{\rho}_{\dot{Y}}(\theta) = \int_0^\infty \left[\frac{1}{\omega}\hat{\rho}_{\dot{X}}\left(\frac{\theta}{\omega}\right)\right] \lambda(\omega) \omega^2 d\omega$$
$$= \int_0^\infty \hat{\rho}_{\dot{X}}\left(\frac{\theta}{\omega}\right) \lambda(\omega) \omega \, d\omega \tag{15.37}$$

($-\infty < \theta < \infty$). As the auto-covariance function of Eq. (15.34) is symmetric, so is the Fourier transform of Eq. (15.37). Hence, Eq. (15.37) implies Eq. (15.10).

15.10.5 Invariance and Eq. (15.13)

Consider Eq. (15.12) and the invariance question following it. As explained in Sect. 15.7, invariance holds if and only if the intensity function of the Poisson process $\mathcal{P}$ is a power function, i.e., $\lambda(\omega) = C\omega^{p-1}$ ($\omega > 0$), where C is a positive coefficient, and where p is a real power. Substituting this power intensity function into Eq. (15.12) yields

$$\hat{\rho}_{\dot{Y}}(\theta) = |\theta|^2 \int_0^\infty \frac{\hat{\rho}_{\dot{X}}(u)}{u^3} \lambda\left(\frac{|\theta|}{u}\right) du$$
$$= |\theta|^2 \int_0^\infty \frac{\hat{\rho}_{\dot{X}}(u)}{u^3} \left[C\left(\frac{|\theta|}{u}\right)^{p-1}\right] du$$
$$= \underbrace{\left[C \int_0^\infty \frac{\hat{\rho}_{\dot{X}}(u)}{u^{2+p}} du\right]}_{c_{\dot{Y}}} \cdot |\theta|^{1+p} \tag{15.38}$$

($-\infty < \theta < \infty$). Evidently—provided that $c_{\dot{Y}}$ is positive—the spectral density of Eq. (15.38) is decreasing in the absolute-value variable $|\theta|$ if and only if the power p is in the range $p < -1$. Setting $p = -\epsilon$ and $C = c\epsilon$ in Eq. (15.38) yields Eq. (15.13).

15.10.6 Equation (15.14)

In terms of the temporal auto-covariance function $\rho_{\dot{Z}}(|\Delta|)$, the temporal variance function of the random motion $Z(t)$ is given by

15.10 Methods

$$v_Z(t) = \text{Var}[Z(t)] = \text{Cov}[Z(t), Z(t)]$$

$$= \text{Cov}\left[\int_0^t \dot{Z}(t_1)\,dt_1, \int_0^t \dot{Z}(t_2)\,dt_2\right]$$

$$= \int_0^t \int_0^t \text{Cov}[\dot{Z}(t_1), \dot{Z}(t_2)]\,dt_1 dt_2 \qquad (15.39)$$

$$= \int_0^t \int_0^t \rho_{\dot{Z}}(|t_1 - t_2|)\,dt_1 dt_2$$

($t \geq 0$). The Fourier inversion formula asserts that

$$\rho_{\dot{Z}}(|\Delta|) = \frac{1}{2\pi}\int_{-\infty}^{\infty} \exp(i\Delta\theta)\,\hat{\rho}_{\dot{Z}}(\theta)\,d\theta \qquad (15.40)$$

($-\infty < \Delta < \infty$). Combining together Eqs. (15.39) and (15.40) yields

$$v_Z(t) = \int_0^t \int_0^t \left\{\frac{1}{2\pi}\int_{-\infty}^{\infty} \exp[i(t_1-t_2)\theta]\,\hat{\rho}_{\dot{Z}}(\theta)\,d\theta\right\} dt_1 dt_2$$

$$= \frac{1}{2\pi}\int_{-\infty}^{\infty}\left\{\left[\int_0^t \exp(i\theta t_1)\,dt_1\right]\left[\int_0^t \exp(-i\theta t_2)\,dt_2\right]\right\}\hat{\rho}_{\dot{Z}}(\theta)\,d\theta \qquad (15.41)$$

($t \geq 0$). Note that

$$\left[\int_0^t \exp(i\theta t_1)\,dt_1\right]\left[\int_0^t \exp(-i\theta t_2)\,dt_2\right]$$

$$= \left[\frac{\exp(i\theta t)-1}{i\theta}\right]\left[\frac{\exp(-i\theta t)-1}{-i\theta}\right] = 2\frac{1-\cos(\theta t)}{\theta^2}. \qquad (15.42)$$

Substituting Eq. (15.42) into Eq. (15.41) yields

$$v_Z(t) = \frac{1}{2\pi}\int_{-\infty}^{\infty}\left\{2\frac{1-\cos(\theta t)}{\theta^2}\right\}\hat{\rho}_{\dot{Z}}(\theta)\,d\theta$$

$$= \frac{2}{\pi}\int_0^{\infty}\frac{1-\cos(\theta t)}{\theta^2}\hat{\rho}_{\dot{Z}}(\theta)\,d\theta \qquad (15.43)$$

$$= \frac{2}{\pi}t\int_0^{\infty}\hat{\rho}_{\dot{Z}}\left(\frac{u}{t}\right)\frac{1-\cos(u)}{u^2}\,du$$

($t \geq 0$); in the transition from the second line of Eq. (15.43) to its last line we used the change-of-variables $\theta \mapsto u = \theta t$. This proves Eq. (15.14).

References

1. I. Eliazar, J. Klafter, Proc. Natl. Acad. Sci. (USA) **106**, 12251 (2009)
2. I. Eliazar, J. Klafter, J. Phys. A: Math. Theor. **42**, 472003 (2009)
3. I. Eliazar, J. Klafter, Phys. Rev. E **82**, 021109 (2010)

4. I. Eliazar, J. Klafter, Ann. Phys. **326**, 2517 (2011)
5. C. Gardiner, *Handbook of Stochastic Methods* (Springer, New York, 2004)
6. N.G. Van Kampen, *Stochastic Processes in Physics and Chemistry*, 3rd edn. (North-Holland, Amsterdam, 2007)
7. Z. Schuss, *Theory and Applications of Stochastic Processes* (Springer, New York, 2009)
8. R. Metzler, J. Klafter, Phys. Rep. **339**, 1 (2000)
9. I.M. Sokolov, J. Klafter, Chaos **15**, 026103 (2005)
10. J. Klafter, I.M. Sokolov, Phys. World **18**, 29 (2005)
11. J.F.C. Kingman, *Poisson Processes* (Oxford University Press, Oxford, 1993)
12. A.N. Shiryaev, *Probability* (Springer, New York, 1995)
13. G. Grimmett, D. Stirzaker, *Probability and Random Processes* (Oxford University Press, Oxford, 2001)
14. G. Lindgren, *Stationary Stochastic Processes* (CRC Press, New York, 2012)
15. M.S. Kesner, Proc. IEEE **70**, 212 (1982)
16. M.B. Weissman, Rev. Mod. Phys. **60**, 537 (1988)
17. B.B. Mandelbrot, *Multifractals and 1/f Noise* (Springer, New York, 1999)
18. D.R. Cox, in *Statistics: An Appraisal*, ed. by H.A. David, H.T. David (Iowa State University Press, Ames, 1984), pp. 55–74
19. P. Doukhan, G. Oppenheim, M.S. Taqqu (eds.), *Theory and Applications of Long-Range Dependence* (Birkhauser, Boston, 2003)
20. G. Rangarajan, M. Ding (eds.), *Processes with Long-Range Correlations: Theory and Applications* (Springer, New York, 2003)
21. W.T. Coffey, Yu.P. Kalmykov, J.T. Waldron, *The Langevin Equation* (World Scientific, Singapore, 2012)
22. G.A. Pavliotis, *Stochastic Processes and Applications* (Springer, New York, 2014)

Chapter 16
First Passage Times

In Chap. 15, we saw how the power Poisson process $\mathcal{E}_-$ emerges when seeking invariance in the context of the aggregates of collections of random motions. In this short chapter, we will show how the power Poisson process $\mathcal{E}_-$ emerges when seeking invariance in the context of the first passage times of collections of random motions. Moreover, we shall see how this invariance quest gives rise to the power Poisson process $\mathcal{E}_+$, and to the Herdan–Heaps law. This chapter is based on [1], which investigated the universal control of random transport processes via Poisson randomizations.

Consider a general space in which general random motions take place. Given an arbitrary "target domain" D within the space, we set our focus on the first passage times (FPTs) to this domain. Namely, for a random motion $Z(t)$ ($t \geq 0$) that takes place in the space, the FPT is the epoch at which the motion reaches the target domain for the very first time:

$$T_Z = \inf\{t \geq 0 \mid Z(t) \in D\}. \tag{16.1}$$

FPTs are of major importance across the sciences [2].

In this chapter, we employ a FPT scheme that is somewhat similar to the aggregation scheme of Chap. 15, and that is described as follows. We set off from a general random motion $X(t)$ ($t \geq 0$) in the space, and generate a countable collection of IID copies $\{X_k(t)\}$ of the random motion. Then, we associate with copy k a time-pace ω_k, where ω_k is a positive parameter, and we run this copy according to the time-pace ω_k; this produces the random motion $Y_k(t) = X_k(\omega_k t)$ ($t \geq 0$) which, in turn, produces the FPT T_{Y_k}. Finally, addressing all the random motions $\{Y_k(t)\}$ collectively, we arrive at the following collection of their FPTs:

$$\mathcal{T} = \{T_{Y_k}\}. \tag{16.2}$$

The FPT definition, combined together with the connection between the random motions $X_k(t)$ and $Y_k(t)$, implies the following connection between the FPTs of these random motions:

$$T_{Y_k} = \inf\{t \geq 0 \mid Y_k(t) \in D\}$$

$$= \inf\{t \geq 0 \mid X_k(\omega_k t) \in D\}$$

$$= \tfrac{1}{\omega_k} \cdot \inf\{t' \geq 0 \mid X_k(t') \in D\} \quad (16.3)$$

$$= \tfrac{1}{\omega_k} \cdot T_{X_k};$$

in the transition from the second line of Eq. (16.3) to its third line we used the change-of-variables $t \mapsto t' = \omega_k t$. The fact that the random motions $\{X_k(t)\}$ are IID copies of the random motion $X(t)$ implies that: the FPTs $\{T_{X_k}\}$ are IID copies of the FPT T_X.

The FPT scheme manifests a mapping whose input is the general random motion $X(t)$, whose output is the collection of FPTs T, and whose parameters are the collection of time-paces $\mathcal{P} = \{\omega_k\}$. As in Chap. 15, we consider the collection $\mathcal{P}$ to be a Poisson process over the positive half-line, with intensity function $\lambda_\mathcal{P}(\omega)$ ($\omega > 0$). Also, we consider the collection $\mathcal{P}$ to be independent of the IID copies $\{X_k(t)\}$ of the random motion $X(t)$. In what follows, we denote by $f_X(t)$ ($t > 0$) the density function of the input's FPT T_X.

Based on the FPT scheme and on the statistics of the collection of time-paces $\mathcal{P}$, a result from the theory of Poisson processes (the "displacement theorem" [3]) asserts that: the collection of FPTs T is a Poisson process over the positive half-line, with intensity function

$$\lambda_T(t) = \frac{1}{t^2} \int_0^\infty \lambda_\mathcal{P}\left(\frac{u}{t}\right) f_X(u) \, u \, du \quad (16.4)$$

($t > 0$); the derivation of Eq. (16.4) is detailed in the Methods.

With Eq. (16.4) at hand, let us address the following *invariance question*: can we set the collection of time-paces $\mathcal{P}$ so that the intensity of the collection of FPTs T be invariant, up to a multiplicative constant, with respect to the input $X(t)$? To answer this question, we seek an intensity function $\lambda_\mathcal{P}(\omega)$ for which $\lambda_T(t) = c_T \cdot \lambda(t)$ ($t \geq 0$), where c_T is a positive coefficient, and where $\lambda(t)$ is an intensity function that does not depend on the input $X(t)$. It is clear from Eq. (16.4) that such invariance holds if and only if the time variable t can be factored out of the term $\lambda_\mathcal{P}(u/t)$, i.e., if and only if the intensity $\lambda_\mathcal{P}(\omega)$ is a power function. Further requiring that the time-paces be bounded from above leads to the following conclusion: invariance holds if and only if the collection of time-paces $\mathcal{P}$ is the power Poisson process $\mathcal{E}_-$; the proof of this conclusion is detailed in the Methods.

Consider the power Poisson process $\mathcal{E}_-$, with its power intensity function $\lambda_-(\omega) = c\epsilon\omega^{-\epsilon-1}$ ($\omega > 0$). Setting $\mathcal{P} = \mathcal{E}_-$, and substituting the power intensity function $\lambda_\mathcal{P}(\omega) = \lambda_-(\omega)$ into Eq. (16.4), yields the following power intensity function:

$$\lambda_T(t) = \underbrace{\left\{c\mathbf{E}\left[T_X^{-\epsilon}\right]\right\}}_{c_T} \cdot \epsilon t^{\epsilon-1} \quad (16.5)$$

16 First Passage Times

($t \geq 0$); the derivation of Eq. (16.5) is detailed in the Methods. Equation (16.5) requires that the input's FPT T_X have a finite moment of order $-\epsilon$.

So, the invariance question gives rise to the power Poisson process $\mathcal{E}_-$—as only $\mathcal{P} = \mathcal{E}_-$ answers this question affirmatively. In turn, it is evident from Eq. (16.5) that setting $\mathcal{P} = \mathcal{E}_-$ implies that $\mathcal{T} = \mathcal{E}_+$, the later equality being in law. Consequently, denoting by $N(t)$ the number of FPTs that occurred up to time t, we obtain the following power temporal mean function:

$$\mathbf{E}[N(t)] = c\mathbf{E}\left[T_X^{-\epsilon}\right] \cdot t^\epsilon \tag{16.6}$$

($t \geq 0$).

Equation (16.5) is a universal design mechanism for FPTs. By setting $\mathcal{P} = \mathcal{E}_-$ and by tuning the parameters of the power Poisson process $\mathcal{E}_-$ (i.e., the coefficient c and the exponent ϵ of its power intensity function), one can produce any desired power intensity function of the output $\mathcal{T} = \mathcal{E}_+$ (the equality being in law). Once a desired exponent ϵ is set, the admissible inputs $X(t)$ are characterized by the moment condition $\mathbf{E}\left[T_X^{-\epsilon}\right] < \infty$. Note that a potentially vast amount of information—the details of the general space, the details of the arbitrary target domain D, and the details of the general random motion $X(t)$—reduces to a single number: $\mathbf{E}\left[T_X^{-\epsilon}\right]$. Moreover, the effect of the number $\mathbf{E}\left[T_X^{-\epsilon}\right]$ on the output can be easily compensated by tuning the coefficient c.

So, the invariance question leads to the implication $\mathcal{P} = \mathcal{E}_- \Rightarrow \mathcal{T} = \mathcal{E}_+$ (the later equality being in law). Based on previous chapters, this implication highlights various connections between the power Poisson processes $\mathcal{E}_-$ and $\mathcal{E}_+$. Here are three such connections (which follow from Chaps. 4 and 7):

- The time-paces that are above a given level are governed by the *Pareto law*; the FPTs that occur up to a given time are governed by the *inverse Pareto law*.
- The fastest time-pace is governed by the *Fréchet law*; the first FPT is governed by the *Weibull law*.
- The order statistics of the time-paces are governed by the *Zipf law*; the order statistics of the FPTs are governed by the *inverse Zipf law*.

Also, the implication $\mathcal{P} = \mathcal{E}_- \Rightarrow \mathcal{T} = \mathcal{E}_+$ highlights a special role of the exponent $\epsilon = 1$. On the one hand, for $\mathcal{P} = \mathcal{E}_-$, the exponent value $\epsilon = 1$ is the only case in which the time-paces are governed by Zipf's law with slope $1/\epsilon = 1$; this unit slope is, empirically, the most commonly encountered Zipf slope [4]. On the other hand, for $\mathcal{T} = \mathcal{E}_+$, the exponent value $\epsilon = 1$ is the only case in which the FPTs form a stationary process—the "standard" Poisson process, which we discussed in Chap. 3. So, somewhat informally, we state that: in order to attain a "stationary flow" of the FPTs $\mathcal{T}$, the time-paces $\mathcal{P}$ must follow Zipf's law with slope one.

We conclude with the empirical *Herdan–Heaps law* [5–7], which is described as follows. Consider reading a very large text from its start. The Herdan–Heaps law asserts that, having read the text's first l words, the number of the different words that we have encountered—the text's running "vocabulary"—is asymptotically approximated by: a monotone increasing and concave power function of the variable l. More broadly, the Herdan–Heaps law is the principal phenomenological form of the *rate of innovation* in general streams of information [8–13].

In the context of the FPT scheme, the innovations are the FPTs $\mathcal{T}$. Consequently, due to the implication $\mathcal{P} = \mathcal{E}_- \Rightarrow \mathcal{T} = \mathcal{E}_+$, the rate of innovation is asymptotically approximated by the power temporal mean function of Eq. (16.6).[1] Thus, for exponents $0 < \epsilon < 1$, Eq. (16.6) manifests a Herdan–Heaps law. Also, for exponents $0 < \epsilon < 1$, the first FPT is governed by the *stretched-exponential law*—which we discussed in the introduction [14–16]. These observations pinpoint a profound connection between the Herdan–Heaps law and the stretched-exponential law—the former originating in linguistics [5–7], and the latter originating in anomalous relaxation [14–16].

Methods

Equation (16.4)

Equations (16.2) and (16.3) imply that

$$\mathcal{T} = \{T_{Y_k}\} = \left\{\frac{1}{\omega_k} \cdot T_{X_k}\right\}. \tag{16.7}$$

Moreover: $\mathcal{P} = \{\omega_k\}$ is a Poisson process over the positive half-line, with intensity function $\lambda_{\mathcal{P}}(\omega)$ ($\omega > 0$); $\{T_{X_k}\}$ are IID copies of the input's FPT T_X; and the Poisson process $\mathcal{P}$ is independent of the FPTs $\{T_{X_k}\}$. Consequently, the "displacement theorem" of the theory of Poisson processes [3] asserts that $\mathcal{T}$ is a Poisson process over the positive half-line, with intensity function

$$\lambda_{\mathcal{T}}(t) = \int_0^\infty \lambda_{\mathcal{P}}(\omega) f_X(\omega t) \omega d\omega \tag{16.8}$$

($t > 0$). The change-of-variables $\omega \mapsto u = \omega t$ leads from Eqs. (16.8) to (16.4).

Invariance and Eq. (16.5)

Consider Eq. (16.4) and the invariance question following it. As explained above, invariance holds if and only if the intensity function of the Poisson process $\mathcal{P}$ is a power function, i.e., $\lambda_{\mathcal{P}}(\omega) = C\omega^{p-1}$ ($\omega > 0$), where C is a positive coefficient, and where p is a real power. Substituting this power intensity function into Eq. (16.4) yields

---

[1] The precise formulation of the asymptotic approximation is described by Eq. (4.3) in Chap. 4.

$$\lambda_T(t) = \tfrac{1}{t^2} \int_0^\infty \lambda_{\mathcal{P}}\left(\tfrac{u}{t}\right) f_X(u)\, u\, du$$

$$= \tfrac{1}{t^2} \int_0^\infty \left[C \left(\tfrac{u}{t}\right)^{p-1} \right] f_X(u)\, u\, du \qquad (16.9)$$

$$= \left[C \int_0^\infty f_X(u)\, u^p du \right] \cdot t^{-p-1}$$

$$= \left\{ C \mathbf{E}\left[T_X^p\right] \right\} \cdot t^{-p-1}$$

($t \geq 0$). The points of the Poisson process $\mathcal{P}$ are bounded from above if and only if the power p is negative. Consequently, setting $p = -\epsilon$ and $C = c\epsilon$ yields

$$\lambda_T(t) = \left\{ c\epsilon \mathbf{E}\left[T_X^{-\epsilon}\right] \right\} \cdot t^{\epsilon-1} \qquad (16.10)$$

($t \geq 0$); this proves Eq. (16.5).

References

1. I. Eliazar, J. Klafter, J. Phys. A Math. Theor. **44**, 222001 (2011)
2. S. Redner, *A Guide to First-Passage Processes* (Cambridge University Press, Cambridge, 2001)
3. J.F.C. Kingman, *Poisson Processes* (Oxford University Press, Oxford, 1993)
4. A. Saichev, D. Sornette, Y. Malevergne, *Theory of Zipf's Law and Beyond* (Springer, New York, 2009)
5. G. Herdan, *Type-Token Mathematic* (Mouton, The Hague, 1960)
6. H.S. Heaps, *Information Retrieval: Computational and Theoretical Aspects* (Academic Press, Boston, 1978)
7. L. Egghe, J. Am. Soc. Inform. Sci. Tech. **58**, 702 (2007)
8. R. Baeza-Yates, G. Navarro, J. Am. Soc. Inform. Sci. **51**, 69 (2000)
9. D.C. van Leijenhorst, T.P. van der Weide, Inform. Sci. **170**, 263 (2005)
10. C. Cattutoa et al., Proc. Natl. Acad. Sci. **106**, 10511 (2009)
11. M.A. Serrano, A. Flammini, F. Menczer, PLoS ONE **4**, e5372 (2009)
12. I. Eliazar, Phys. A **390**, 3189 (2011)
13. I. Eliazar, M.H. Cohen, Ann. Phys. **332**, 56 (2012)
14. G. Williams, D.C. Watts, Trans. Faraday Soc. **66**, 80 (1970)
15. J.C. Phillips, Rep. Prog. Phys. **59**, 1133 (1996)
16. W.T. Coffey, Y.P. Kalmykov, *Fractals, Diffusions and Relaxation in Disordered Systems* (Wiley, New York, 2006)

Chapter 17
From Power to Lognormal

In Chap. 2, we set off from lognormal statistics and arrived at the power Poisson processes $\mathcal{E}_+$ and $\mathcal{E}_-$. In this section, we "close the circle": we set off from the power Poisson processes $\mathcal{E}_+$ and $\mathcal{E}_-$ and return back to our starting point—lognormal statistics.

17.1 Double-Pareto Laws

In his seminal work, the French economist Thomas Piketty asserted—based on vast empirical evidence—that the rate of return on capital is greater than the rate of economic growth [1]. Piketty's empirical findings imply that the financial evolution of the rich diverges from the financial evolution of the proletariat, thus leading to the concentration of wealth. In this section, we address Piketty's thesis from the perspective of Chap. 2.

As noted in Chap. 2, the stochastic differential equation appearing in Sect. 2.2 of that chapter is a paradigmatic model of evolution in economics and finance. This model yields Geometric Brownian Motion (GBM), and it is has two parameters: the "drift" r, and the "volatility" $\sqrt{\nu}$. Motivated by Piketty's thesis, we consider two GBMs. The first GBM governs the financial evolution of the proletariat, and it is assumed to have a low return rate (r) and a low-risk level ($\sqrt{\nu}$) such that: $r > \frac{1}{2}\nu$. The second GBM governs the financial evolution of the rich, and it is assumed to have a high return rate (r) and a high-risk level ($\sqrt{\nu}$), such that: $r < \frac{1}{2}\nu$.

More generally, we can consider the financial evolutions of the proletariat and of the rich to be driven by two Gaussian motions (GMs)—as described in Sects. 2.4 and 2.5 of Chap. 2. In this general setting, the GM driving the financial evolution of the proletariat is assumed to have a positive limiting mean-to-variance ratio: Eq. (2.10) of Chap. 2 with $p > 0$. And, the GM driving the financial evolution of the rich is assumed to have a negative limiting mean-to-variance ratio: Eq. (2.10) of Chap. 2 with $p < 0$. The aforementioned double GBM setting is a special case of this double GM setting.

© Springer Nature Switzerland AG 2020
I. Eliazar, *Power Laws*, Understanding Complex Systems,
https://doi.org/10.1007/978-3-030-33235-8_17

The financial evolution of the proletariat leads, via the Poisson-process limit of Chap. 2, to a limiting Poisson process $\mathcal{E}_+$ with the power intensity function $\lambda_+(x) = c_+\epsilon_+ x^{\epsilon_+ - 1}$ ($x > 0$); the coefficient c_+ and the exponent ϵ_+ are positive parameters. The financial evolution of the rich leads, via the Poisson-process limit of Chap. 2, to a limiting Poisson process $\mathcal{E}_-$ with the power intensity function $\lambda_-(x) = c_-\epsilon_- x^{-\epsilon_- - 1}$ ($x > 0$); the coefficient c_- and the exponent ϵ_- are positive parameters. As the limiting Poisson process $\mathcal{E}_+$ manifests the wealth values of the proletariat, it applies up to some positive threshold level l_+. And, as the limiting Poisson process $\mathcal{E}_-$ manifests the wealth values of the rich, it applies above some positive threshold level l_-.

Considering the threshold level l_+ to be smaller than the threshold level l_-, and merging the two limiting Poisson processes $\mathcal{E}_+$ and $\mathcal{E}_-$ together, we arrive at a joint Poisson process $\mathcal{E}_\pm$ whose intensity function $\lambda_\pm(x)$ ($x > 0$) admits the following three-part structure. (I) Proletariat part: $\lambda_\pm(x) = c_+\epsilon_+ x^{\epsilon_+ - 1}$ for $0 < x < l_+$. (II) Rich part: $\lambda_\pm(x) = c_-\epsilon_- x^{-\epsilon_- - 1}$ for $l_- < x < \infty$. (III) Middle-class part: an intermediate part which spans the range $l_+ \leq x \leq l_-$, and connects the proletariat part to the rich part.

The power intensity function $\lambda_+(x)$ is integrable at the origin (but not at infinity), and the power intensity function $\lambda_-(x)$ is integrable at infinity (but not at the origin). Consequently, the "amalgam" intensity function $\lambda_\pm(x)$ is integrable over the entire positive half-line: $m_\pm = \int_0^\infty \lambda_\pm(x)\,dx < \infty$. This integrability further implies that the "amalgam" Poisson process $\mathcal{E}_\pm$ comprises finitely many points [2]. In turn, a general result from the theory of Poisson processes asserts that these finitely many points form a collection of IID random variables [2].

The common probability density function of these IID random variables, $f_\pm(x)$ ($x > 0$), is given by the unit-mass normalization of the "amalgam" intensity function $\lambda_\pm(x)$. Namely [2]: $f_\pm(x) = \lambda_\pm(x)/m_\pm$ ($x > 0$). Consequently, we have

$$f_\pm(x) = \frac{c_+}{m_\pm} \epsilon_+ x^{\epsilon_+ - 1} \tag{17.1}$$

for $0 < x < l_+$, and

$$f_\pm(x) = \frac{c_-}{m_\pm} \epsilon_- x^{-\epsilon_- - 1} \tag{17.2}$$

for $l_- < x < \infty$.

The proletariat part of the density function $f_\pm(x)$—given by Eq. (17.1)—manifests an *inverse-Pareto law* with exponent ϵ_+. The rich part of the density function $f_\pm(x)$—given by Eq. (17.2)—manifests a *Pareto law* with exponent ϵ_-. Probability density functions with the amalgamated inverse-Pareto and Pareto structure of Eqs. (17.1)–(17.2) characterize the *double-Pareto law* [3–8]. This law is encountered prevalently across the sciences [9–22].

On the one hand, the proletariat part of the density function $f_\pm(x)$ is monotone decreasing in the exponent range $\epsilon_+ < 1$, and is monotone increasing in the exponent range $\epsilon_+ > 1$. On the other hand, the rich part of the density function $f_\pm(x)$ is always monotone decreasing. Hence, the *simplest shape* of the density function $f_\pm(x)$ is

17.1 Double-Pareto Laws 185

as follows: monotone decreasing for exponents $\epsilon_+ < 1$, and unimodal for exponents $\epsilon_+ > 1$. Based on [23] and [24], in the remainder of this chapter we shall construct a mechanism that generates density functions $f_\pm(x)$ with such a shape.

17.2 Langevin and Gibbs

Consider a motion that takes place over the real line, and denote by $Z(t)$ ($t \geq 0$) the motion's position at time t. The motion follows gradient-descent dynamics over an "energy landscape" that is quantified by the *potential function* $V(z)$ ($-\infty < z < \infty$). Specifically, the motion's dynamics are governed by the nonlinear ordinary differential equation

$$\dot{Z}(t) = F[Z(t)] \qquad (17.3)$$

($t \geq 0$), where $F(z)$ is the corresponding *force function*: the negative gradient $F(z) = -V'(z)$ ($-\infty < z < \infty$) of the potential function.

As noted in Chap. 2, in the real-world noise is omnipresent and always present. Adding noise to the motion leads us from the nonlinear ODE of Eq. (17.3) to the following stochastic differential equation (SDE) [25–27]:

$$\dot{Z}(t) = F[Z(t)] + \sqrt{\nu} \cdot \eta(t) \qquad (17.4)$$

($t \geq 0$), where ν is a positive parameter, and where $\eta(t)$ ($t \geq 0$) is a noise process manifesting random fluctuations.

As in the SDE appearing in Sect. 2.2 of Chap. 2, the parameter $\sqrt{\nu}$ is the noise magnitude, and $\eta(t)$ is the noise level at time t. The SDE (17.4)—termed *Langevin equation* [28], in honor of Paul Langevin [29]—is a foundational in the physical sciences [30, 31]. As in Chap. 2, we consider the noise process $\eta(t)$ to be *white noise* [30, 31].

Provided that the underlying potential function $V(z)$ is 'well behaved', the Langevin equation (17.4) admits a *steady state*. A real-valued random variable Z_* is a steady state of the Langevin equation (17.4) if the following implication holds: $Z(0) = Z_* \Rightarrow Z(t) = Z_*$ for all $t > 0$, the equalities being in law. Namely, if we initiate the motion from a position that is the random variable Z_* then, at any future time t, the motion's position $Z(t)$ will be statistically identical to its initial position Z_*.

The statistical distribution of the steady-state random variable Z_* is given by the following probability density function:

$$\frac{d}{dz} \Pr(Z_* < z) = c_* \exp\left[-\frac{1}{\tau} V(z)\right] \qquad (17.5)$$

($-\infty < z < \infty$), where c_* is a normalizing constant, and where $\tau = \nu/2$. The density function of Eq. (17.5) is termed *Gibbs density*, and the statistical-physics interpretation of the parameter τ is *temperature* [32]. The derivation of the Gibbs density of Eq. (17.5) is obtained via the *Fokker–Planck equation* that corresponds to the Langevin equation (17.4) [33].

The Gibbs density of Eq. (17.5) also emerges from an altogether different approach: the Jaynes principle of *entropy maximization* [34–37]. Indeed, assume that we are seeking a random variable Z_* whose entropy is maximal, subject to the following constraint: $E[V(Z_*)] = v_*$, where v_* is a real value that is within the range of the potential function $V(z)$. Then, the entropy-maximizing random variable Z_* is governed by the Gibbs density of Eq. (17.5), where the inverse temperature $1/\tau$ is a Lagrange multiplier emanating from the entropy-maximization optimization problem.

17.3 Exponentiation

As established in Chap. 2, Geometric Brownian Motion (GBM) universally leads—via the Poisson-process limit of that chapter—to the power Poisson processes $\mathcal{E}_+$ and $\mathcal{E}_-$. Also, as established in Chap. 13, power evolution schemes universally lead—via the limit laws presented in Sects. 13.4 and 13.5 of that chapter—to the power Poisson processes $\mathcal{E}_+$ and $\mathcal{E}_-$. The GBM stochastic dynamics are governed by the SDE (2.3) of Chap. 2, and the deterministic dynamics of the power evolution schemes are governed by the ODE (12.2) of Chap. 12.

Both the SDE (2.3) of Chap. 2 and the ODE (12.2) of Chap. 12 describe *multiplicative dynamics* over the positive half-line. Specifically, in both the SDE and the ODE $X(t)$ denotes a positive size (at time t), and the rate of change is measured relatively to the evolving size: $\dot{X}(t)/X(t)$, as appearing in the left-hand sides of the SDE and the ODE. On the other hand, the Langevin equation (17.4) describes *additive dynamics* over the real line. Specifically, in the Langevin equation (17.4), $Z(t)$ denotes a real position (at time t), and the rate of change is measured in an absolute fashion: $\dot{Z}(t)$, as appearing in the left-hand side of Eq. (17.4).

With regard to the real position $Z(t)$ ($t \geq 0$) of Sect. 17.2, consider now the following exponential transformation: $Z(t) \mapsto X(t) = \exp[Z(t)]$ ($t \geq 0$). This exponential transformation shifts us from the real position $Z(t)$ to the positive size $X(t)$.

Applying Ito's formula [38–40], this exponential transformation shifts us from the Langevin equation (17.4) to the following *geometric Langevin equation*:

$$\frac{\dot{X}(t)}{X(t)} = F_G[X(t)] + \sqrt{\nu} \cdot \eta(t) \tag{17.6}$$

($t \geq 0$), where

$$F_G(x) = \tau + F[\ln(x)] \tag{17.7}$$

($0 < x < \infty$) is the corresponding "*geometric force function*", and where the noise magnitude $\sqrt{\nu}$ and the noise level $\eta(t)$ are as in the Langevin equation (17.4).

17.3 Exponentiation

The exponential transformation—which shifts us from the Langevin equation (17.4) to the geometric Langevin equation (17.6)—is further induced to the steady states of these equations. Indeed, with regard to the steady state Z_* of the Langevin equation (17.4), apply the exponential transformation: $Z_* \mapsto X_* = \exp(Z_*)$. Then, the random variable X_* is the steady state of the geometric Langevin equation (17.6): $X(0) = X_* \Rightarrow X(t) = X_*$ for all $t > 0$, the equalities being in law. Namely, if we initiate the geometric Langevin equation (17.6) from a size that is the random variable X_* then, at any future time t, the size $X(t)$ will be statistically identical to the initial size X_*.

The statistical distribution of the steady-state random variable X_* is given by the following *log-Gibbs density*:

$$\frac{d}{dx} \Pr(X_* < x) = c_* \exp\left[-\frac{1}{\tau} V_G(x)\right] \tag{17.8}$$

$(0 < x < \infty)$, where

$$V_G(x) = \tau \ln(x) + V[\ln(x)] \tag{17.9}$$

$(0 < x < \infty)$ is the corresponding "*geometric potential function*", and where the normalizing constant c_* and the temperature parameter τ are as in the Gibbs density of Eq. (17.5).

The log-Gibbs density of Eq. (17.8) can be derived in several ways. One way is to obtain it via the *Fokker–Planck equation* that corresponds to the geometric Langevin equation (17.6) [33]. Another way is to obtain it from the Gibbs density of Eq. (17.5) via the exponential transformation $Z_* \mapsto X_* = \exp(Z_*)$; this derivation is detailed in the Methods. The log-Gibbs density of Eq. (17.8) emerges from entropy maximization on a *logarithmic scale*: for a given positive-valued random variable X_*, apply the entropy maximization described at the end of Sect. 17.2 to its logarithm—the real-valued random variable $Z_* = \ln(X_*)$.

17.4 U-Shaped Potentials

Consider now *U-shaped* potentials. Specifically, consider the potential function $V(z)$ to be convex and smooth, $V''(z) > 0$, and to have a global minimum. With no loss of generality, we assume that the global minimum is attained at the origin: $V(z) > V(0)$ for all $z \neq 0$. Also, we denote by $G(z) = V'(z)$ $(-\infty < z < \infty)$ the gradient of the potential function $V(z)$.

The U-shape of the potential function implies that its gradient $G(z)$ is a monotone increasing function, $G'(z) > 0$, and that it passes through the origin, $G(0) = 0$. In turn, the force function of the Langevin equation (17.4), $F(z) = -G(z)$ $(-\infty < z < \infty)$, "pushes" toward the origin: $F(z)$ is positive over the negative half-line, and is negative over the positive half-line. Hence, denoting by z_{min} the level at which the global minimum of the potential function $V(z)$ is attained, and denoting by z_{push}

the level toward which the force function $F(z)$ is pushing, we arrive at

$$z_{min} = 0 = z_{push}. \tag{17.10}$$

The exponential transformation of Sect. 17.3 changes matters dramatically. This change is manifested by the potential function $V(z)$ versus the geometric potential function $V_G(x)$, and by the force function $F(z)$ versus the geometric force function $F_G(x)$, as we shall now argue. In what follows, $G^{-1}(y)$ denotes the inverse function of the gradient $G(z)$.[1]

As set, the potential function $V(z)$ is U-shaped, and its global minimum is attained at the origin: $z_{min} = 0$. On the other hand, the geometric potential function $V_G(x)$ of Eq. (17.9) admits two different shapes, depending on the interplay between the temperature τ and the gradient's lower bound $G(-\infty) = \lim_{z \to -\infty} G(z)$. (I) For temperatures $\tau < -G(-\infty)$, the geometric potential function $V_G(x)$ is U-shaped, and its global minimum is attained at the positive level $x_{min} = \exp\left[G^{-1}(-\tau)\right]$. (II) If the gradient $G(z)$ is bounded from below then: for temperatures $\tau \geq -G(-\infty)$ the geometric potential function $V_G(x)$ is monotone increasing, and hence its global minimum is attained at the zero level $x_{min} = 0$.

As noted above, the force function $F(z)$ pushes toward the origin: $z_{push} = 0$. On the other hand, the geometric force function $F_G(x)$ of Eq. (17.7) displays two different scenarios, depending on the interplay between the temperature τ and the gradient's upper bound $G(\infty) = \lim_{z \to \infty} G(z)$. (I) For temperatures $\tau < G(\infty)$, the geometric force function $F_G(x)$ "pushes" toward the positive level $x_{push} = \exp\left[G^{-1}(\tau)\right]$—as $F_G(x)$ is positive below the level x_{push}, and as $F_G(x)$ is negative above level x_{push}. (II) If the gradient $G(z)$ is bounded from above then: for temperatures $\tau \geq G(\infty)$ the geometric force function $F_G(x)$ "pushes" toward the infinite level $x_{push} = \infty$—as $F_G(x)$ is positive over the entire positive half-line.

The aforementioned formulae for the levels x_{min} and x_{push}, combined together with the properties of the gradient $G(z)$, imply that

$$0 \leq x_{min} < 1 < x_{push} \leq \infty. \tag{17.11}$$

Equation (17.11) manifests the dramatic effect of the exponential transformation of Sect. 17.3. On the one hand, in Eq. (17.10) there is no daylight between the levels z_{min} and z_{push}. On the hand, in Eq. (17.11) there is always a rift between the levels x_{min} and x_{push}, and this rift may even be infinitely large. Indeed, for bounded gradients and high temperatures we have $x_{min} = 0$ and $x_{push} = \infty$.

The rift between the levels x_{min} and x_{push} can be quantified by the following "rift-score":

$$\mathcal{R} = 1 - \frac{x_{min}}{x_{push}}. \tag{17.12}$$

---

[1] The gradient is a monotone increasing function, and hence its inverse function is well-defined indeed.

17.4 U-Shaped Potentials

The rift-score $\mathcal{R}$ is an *inequality index* [41, 42] that quantifies the inherent socioeconomic inequality of the steady-state random variable X_*.[2] Being an inequality index, the rift-score takes values in the unit interval, $0 \leq \mathcal{R} \leq 1$. The rift-score meets its zero lower bound if and only if the levels x_{min} and x_{push} coincide: $\mathcal{R} = 0 \Leftrightarrow x_{min} = 1 = x_{push}$. Also, the rift-score meets its unit upper bound if and only if either the level x_{min} is zero or the level x_{push} is infinite: $\mathcal{R} = 1 \Leftrightarrow x_{min} = 0$ or $x_{push} = \infty$.

The rift-score $\mathcal{R}$ is a monotone increasing function of the temperature τ. As such, the rift-score $\mathcal{R}$ involves the critical temperature $\tau_c = \min\{-G(-\infty), G(\infty)\}$, and it admits the following features. (I) It meets its zero lower bound if and only if the temperature drops to zero: $\mathcal{R} \to 0 \Leftrightarrow \tau \to 0$. (II) For temperatures smaller than the critical temperature, $\tau < \tau_c$, it is given by

$$\mathcal{R} = 1 - \exp\left[G^{-1}(-\tau) - G^{-1}(\tau)\right]. \tag{17.13}$$

(III) It meets its unit upper bound if and only if the temperature reaches or exceeds the critical temperature: $\mathcal{R} = 1 \Leftrightarrow \tau \geq \tau_c$.

17.5 Edge of Convexity

As in the previous section, in this section we also consider the potential function $V(z)$ to be U-shaped: convex, smooth, and with a global minimum attained at the origin. With such a potential function, the Gibbs density of Eq. (17.5) is *unimodal*: monotone increasing over the negative half-line ($z < 0$), attaining its global maximum at the origin ($z = 0$), and monotone decreasing over the positive half-line ($z > 0$).

The Gibbs density manifests a statistical balance between two opposites of the Langevin equation (17.4): on the one hand the deterministic force $F(z)$, which pushes the motion toward the origin; on the other hand, the stochastic white noise $\eta(t)$, which drives the motion away from the origin. The temperature τ plays a key role in this statistical balance: the smaller the temperature—the more dominant the "deterministic hand", and the more pronounced the *mode* of the Gibbs density; the larger the temperature—the more dominant the "stochastic hand", and the more pronounced the *tails* of the Gibbs density. However, no matter how small or how large the temperature τ is—the Gibbs density always maintains its unimodal shape, and its mode is always the origin: $\mathbf{M}[Z_*] = 0$.

In Sect. 17.4, we saw that the exponential transformation of Sect. 17.3 changes matters dramatically. In this section, we examine the effect of the exponential transformation on the log-Gibbs density of Eq. (17.8). To that end, we focus on U-shaped potentials that are on the "*edge of convexity*" [23, 24]: the potential function $V(z)$ has affine asymptotes in the limits $z \to -\infty$ and $z \to +\infty$. In turn, these asymptotes of the potential function $V(z)$ imply that its gradient $G(z)$ is bounded: $-\infty < G(-\infty) < 0 < G(\infty) < \infty$.

---

[2] Recall that we described and used inequality indices in Chap. 9.

On the "edge of convexity" the log-Gibbs density of Eq. (17.8) displays the following *power asymptotics*:

$$\frac{d}{dx} \Pr(X_* < x) \approx \begin{cases} x^{\epsilon_+ - 1} & (\text{as } x \to 0), \\ x^{-\epsilon_- - 1} & (\text{as } x \to \infty), \end{cases} \qquad (17.14)$$

with exponents

$$\epsilon_+ = -\frac{1}{\tau} G(-\infty) \text{ and } \epsilon_- = \frac{1}{\tau} G(\infty). \qquad (17.15)$$

The derivation of Eqs. (17.14) and (17.15) is detailed in the Methods.

So, on the "edge of convexity" the log-Gibbs density admits three different shapes, depending on the exponent ϵ_+. (I) For exponent values $0 < \epsilon_+ < 1$, the density is monotone decreasing and unbounded, as it explodes near the origin. (II) At the exponent value $\epsilon_+ = 1$, the density is monotone decreasing and bounded. (III) For exponent values $1 < \epsilon_+ < \infty$, the density is unimodal. The mode of the log-Gibbs density coincides with the level at which the geometric potential function $V_G(x)$ attains its global minimum: $\mathbf{M}[X_*] = x_{min}$. There are two different scenarios for the mean of the log-Gibbs density, depending on the value of the exponent ϵ_-. (I) For exponent values $\epsilon_- \leq 1$ the mean diverges, $\mathbf{E}[X_*] = \infty$. (II) For exponent values $\epsilon_- > 1$, the mean converges, $\mathbf{E}[X_*] < \infty$.

The behavior of the log-Gibbs density is determined by the interplay between the temperature τ and the bounds of the gradient $G(z)$, via a pair of *phase transitions*. (I) As the temperature τ exceeds the level $-G(-\infty)$, the *mode* of the log-Gibbs density *vanishes*. (II) As the temperature τ exceeds the level $G(\infty)$, the *mean* of the log-Gibbs density *explodes*. Thus, the log-Gibbs density displays two diametric behaviors: *"normal"* and *"anomalous"*. The normal behavior holds for low temperatures, the anomalous behavior holds for high temperatures, and Table 17.1 summarizes the key features of these two behaviors.

Table 17.1 The "normal" and "anomalous" behaviors of the log-Gibbs density of Sect. 17.5, resulting from U-shaped potentials that are on the "edge of convexity". First and second lines: the temperatures and the exponents characterizing these behaviors; the critical temperature is $\tau_c = \min\{-G(-\infty), G(\infty)\}$ (with $-\infty < G(-\infty) < 0 < G(\infty) < \infty$). Third and fourth lines: the density's shape and values. Fifth and sixth lines: the density's mode and mean

	Normal	Anomalous
Temperature	$\tau < \tau_c$	$\tau > \tau_c$
Exponents	$1 < \epsilon_+, \epsilon_- < \infty$	$0 < \epsilon_+, \epsilon_- < 1$
Shape	unimodal	monotone decreasing
Values	bounded	unbounded
Mode	$0 < \mathbf{M}[X_*] < \infty$	$\mathbf{M}[X_*] = 0$
Mean	$0 < \mathbf{E}[X_*] < \infty$	$\mathbf{E}[X_*] = \infty$

17.5 Edge of Convexity

From a socioeconomic perspective, the normal behavior and the anomalous behavior describe two diametric societies. (I) A "normal" society comprising a small poor class, a large middle class, and a small rich class; in this society the rift between the poor and the rich is finite, $0 < x_{min} < x_{push} < \infty$. (II) A "polar" society comprising a vast poor class and a small super-rich class; in this society, there is no middle class, and the rift between the poor and the super-rich is infinite, $0 = x_{min} < x_{push} = \infty$.

17.6 Universal Approximation

As in Sect. 17.4, in this section, we consider the potential function $V(z)$ to be U-shaped: convex, smooth, and with a global minimum attained at the origin. Also, as in Sect. 17.5, in this section we consider the gradient $G(z)$ to be bounded: $-\infty < G(-\infty) < 0 < G(\infty) < \infty$.

The gradient $G(z)$ admits the following *sigmoidal shape* over the real line: monotone increasing from the lower bound $G(-\infty)$ to the upper bound $G(\infty)$, and passing through the origin, $G(0) = 0$. The slope of the gradient $G(z)$ at the origin is positive: $G'(0) > 0$. Due to its sigmoidal shape, the gradient $G(z)$ admits the following *piecewise affine approximation*:

$$G_{\pm}(z) = \begin{cases} G(-\infty) & -\infty < z < \frac{G(-\infty)}{G'(0)}, \\ G'(0) \cdot z & \frac{G(-\infty)}{G'(0)} \leq z \leq \frac{G(\infty)}{G'(0)}, \\ G(\infty) & \frac{G(\infty)}{G'(0)} < z < \infty. \end{cases} \quad (17.16)$$

The approximate gradient $G_{\pm}(z)$ is a continuous function over the real line that: is constant below the negative level $G(-\infty)/G'(0)$; is linear between the negative level $G(-\infty)/G'(0)$ and the positive level $G(\infty)/G'(0)$; and is constant above the positive level $G(\infty)/G'(0)$. The reason why we denote the approximate gradient $G_{\pm}(z)$ will become apparent soon.

The approximate gradient $G_{\pm}(z)$ yields an approximate potential function $V_{\pm}(z)$ that satisfies: $V'_{\pm}(z) = G_{\pm}(z)$ ($-\infty < z < \infty$). In turn, the approximate potential function $V_{\pm}(z)$ yields a Gibbs density via Eq. (17.5), and a log-Gibbs density via Eq. (17.8). Using the notation of Eq. (17.15), as well as the notation $\sigma^2 = \tau/G'(0)$, the resulting log-Gibbs density is

$$f_{\pm}(x) = \frac{c_*}{x} \cdot \begin{cases} \exp\left(\frac{1}{2}\sigma^2\epsilon_+^2\right) \cdot x^{\epsilon_+} & 0 < x < \exp\left(-\sigma^2\epsilon_+\right), \\ \exp\left[-\frac{1}{2\sigma^2}\ln(x)^2\right] & \exp\left(-\sigma^2\epsilon_+\right) \leq x \leq \exp\left(\sigma^2\epsilon_-\right), \\ \exp\left(\frac{1}{2}\sigma^2\epsilon_-^2\right) \cdot x^{-\epsilon_-} & \exp\left(\sigma^2\epsilon_-\right) < x < \infty, \end{cases} \quad (17.17)$$

where c_* is a normalizing constant.

The log-Gibbs density of Eq. (17.17) comprises three parts. (I) Left part: below the threshold level $l_+ = \exp(-\sigma^2 \epsilon_+)$ the density manifests an *inverse-Pareto* form with exponent ϵ_+. (II) Right part: above the threshold level $l_- = \exp(\sigma^2 \epsilon_-)$ the density manifests a *Pareto* form with exponent ϵ_-. (III) Center part: in between the threshold levels $l_+ = \exp(-\sigma^2 \epsilon_+)$ and $l_- = \exp(\sigma^2 \epsilon_-)$ the density manifests a *lognormal* form with normal variance σ^2 [43, 44]; we shall elaborate on the lognormal law in the next section.

17.7 Lognormal and Log-Laplace Scenarios

Consider the approximated gradient of Eq. (17.16), and set its bounds to be infinite: $G(-\infty) = -\infty$ and $G(\infty) = \infty$. This yields an approximate gradient $G_\pm(z)$ that is a *linear* function: $G_\pm(z) = G'(0) \cdot z$ ($-\infty < z < \infty$). The Gibbs density resulting from the linear approximate gradient is *normal (Gauss)* [45]:

$$\frac{d}{dz} \Pr(Z_* < z) = \frac{1}{\sqrt{2\pi\sigma^2}} \exp\left(-\frac{z^2}{2\sigma^2}\right) \tag{17.18}$$

($-\infty < z < \infty$), where σ^2 is the corresponding variance.

In turn, the log-Gibbs density resulting from the linear approximate gradient is *lognormal* [43, 44]:

$$f_\pm(x) = \frac{1}{\sqrt{2\pi\sigma^2}} \frac{1}{x} \exp\left[-\frac{\ln(x)^2}{2\sigma^2}\right] \tag{17.19}$$

($0 < x < \infty$). The lognormal density of Eq. (17.19) manifests the following scenario regarding the log-Gibbs density of Eq. (17.17): the center part of the density takes over the entire positive half-line; this "takeover" eliminates the left inverse-Pareto part and the right Pareto part.

Consider the approximated gradient of Eq. (17.16), and set its slope at the origin to be infinite: $G'(0) = \infty$. This yields an approximate gradient $G_\pm(z)$ that is a *Heaviside* function: $G_\pm(z) = G(-\infty)$ over the negative half-line ($z < 0$), and $G_\pm(z) = G(\infty)$ over the positive half-line ($z > 0$). The Gibbs density resulting from the Heaviside approximate gradient is *Laplace* [46]:

$$\frac{d}{dz} \Pr(Z_* < z) = \frac{\epsilon_+ \epsilon_-}{\epsilon_+ + \epsilon_-} \cdot \begin{cases} \exp(\epsilon_+ z) & -\infty < z \leq 0, \\ \exp(-\epsilon_- z) & 0 \leq z < \infty. \end{cases} \tag{17.20}$$

Consequently, the log-Gibbs density resulting from the Heaviside approximate gradient is *log-Laplace* [46]:

17.7 Lognormal and Log-Laplace Scenarios

$$f_{\pm}(x) = \frac{\epsilon_+ \epsilon_-}{\epsilon_+ + \epsilon_-} \cdot \begin{cases} x^{\epsilon_+ - 1} & 0 < x \leq 1, \\ x^{-\epsilon_- - 1} & 1 \leq x < \infty. \end{cases} \quad (17.21)$$

The log-Laplace density of Eq. (17.21) manifests the following scenario regarding the log-Gibbs density of Eq. (17.17): the left inverse-Pareto part of the density takes over up to the unit level, and the right Pareto part of the density takes over down to the unit level; this "takeover" eliminates the center lognormal part.

17.8 Summary

Piketty's empirical findings, when combined together with the geometric evolution models of Chap. 2, lead to the merger of the power Poisson processes $\mathcal{E}_+$ and $\mathcal{E}_-$. In turn, this merger yields the double-Pareto law: a statistical distribution whose small "poor" values are governed by the inverse-Pareto law, and whose large "rich" values are governed by the Pareto law. The simplest shape for the probability density function of the double-Pareto law is—depending on its inverse-Pareto exponent— either monotone decreasing or unimodal.

The geometric Langevin equation is a mechanism that generates the double-Pareto law with the aforementioned simplest shape. Indeed, consider the potential function underpinning the geometric Langevin equation to be on the "edge of convexity": convex, U-shaped, and with affine asymptotes. Then, the resulting steady state is double-Pareto, and the shape of its log-Gibbs density is—depending on the temperature parameter of the geometric Langevin equation—either monotone decreasing or unimodal.

The shape of the gradient function of a potential function that is on the "edge of convexity" is sigmoidal: monotone increasing from a negative lower bound to a positive upper bound. In turn, the piecewise affine approximation of a sigmoidal gradient function yields a universal approximation for the corresponding log-Gibbs density. This universal approximation comprises three parts: (I) small "poor" values that are governed by inverse-Pareto statistics; (II) large "rich" values that are governed by Pareto statistics; and (III) intermediate "middle-class" values that are governed by lognormal statistics. The approximation is universal in the sense that, in essence, it does not depend on the details of the underlying potential function.

Thus, the universal approximation implies that lognormal statistics are a "universal bridge" connecting the inverse-Pareto part and the Pareto part of the double-Pareto law. Again: the universality of the "lognormal bridge" is in the sense that it does not depend on the details of the underlying potential function (which is on the "edge of convexity"). As the double-Pareto law emerged from the merger of the power Poisson processes $\mathcal{E}_+$ and $\mathcal{E}_-$, we thus "closed the circle". Indeed, in Chap. 2 we set off from lognormal statistics and arrived at the power Poisson processes $\mathcal{E}_+$ and $\mathcal{E}_-$. On the other hand, in this chapter we set off from the power Poisson processes $\mathcal{E}_+$ and $\mathcal{E}_-$ and returned back to our starting point: lognormal statistics—which emerge as the aforementioned "universal bridge".

17.9 Methods

17.9.1 Equation (17.8)

The random variable X_* is the exponentiation of the real-valued random variable Z_*, i.e., $X_* = \exp(Z_*)$. Hence, we have

$$\Pr(X_* < x) = \Pr\left[\exp(Z_*) < x\right] = \Pr[Z_* < \ln(x)] \qquad (17.22)$$

($0 < x < \infty$). Differentiating both sides of Eq. (17.22) yields

$$\frac{d}{dx}\Pr(X_* < x) = \left(\frac{d}{dz}\Pr(Z_* < z)\right)\bigg|_{z=\ln(x)} \cdot \frac{1}{x} \qquad (17.23)$$

($0 < x < \infty$). Substituting the Gibbs density of Eq. (17.5) into Eq. (17.23) further yields

$$\begin{aligned}\frac{d}{dx}\Pr(X_* < x) &= c_* \exp\left\{-\frac{1}{\tau}V[\ln(x)]\right\} \cdot \frac{1}{x} \\ &= c_* \exp\left\{-\frac{1}{\tau}V[\ln(x)] - \ln(x)\right\}\end{aligned} \qquad (17.24)$$

($0 < x < \infty$). Introduce the following "geometric potential function":

$$V_G(x) = \tau \ln(x) + V[\ln(x)] \qquad (17.25)$$

($0 < x < \infty$). Combining together Eqs. (17.24) and (17.25), we obtain the following representation of the log-Gibbs density:

$$\frac{d}{dx}\Pr(X_* < x) = c_* \exp\left[-\frac{1}{\tau}V_G(x)\right] \qquad (17.26)$$

($0 < x < \infty$); this proves Eq. (17.8).

17.9.2 Equations (17.14) and (17.15)

On the "edge of convexity", the U-shaped potential function $V(z)$ has affine asymptotes in the limits $z \to -\infty$ and $z \to +\infty$. Namely, in the left limit we have

$$\lim_{z \to -\infty}\{V(z) - [a_- z + b_-]\} = 0, \qquad (17.27)$$

17.9 Methods

where a_- and b_- are the parameters of the left asymptote. And, in the right limit we have

$$\lim_{z \to +\infty} \{V(z) - [a_+ z + b_+]\} = 0, \tag{17.28}$$

where a_+ and b_+ are the parameters of the right asymptote.

Using the gradient $G(z) = V'(z)$ of the potential function, Eq. (17.27) can be expressed in the following integral form:

$$V(0) + \int_{-\infty}^{0} [a_- - G(u)] \, du = b_- . \tag{17.29}$$

From Eq. (17.29), it is evident that $a_- = G(-\infty)$. Also, using the gradient $G(z) = V'(z)$ of the potential function, Eq. (17.28) can be expressed in the following integral form:

$$V(0) + \int_{0}^{\infty} [G(u) - a_+] \, du = b_+ . \tag{17.30}$$

From Eq. (17.30), it is evident that $a_+ = G(\infty)$.

Equation (17.24) implies that the log-Gibbs density can be written as follows:

$$c_* \exp\left\{-\frac{1}{\tau} V[\ln(x)]\right\} \cdot \frac{1}{x} \tag{17.31}$$

($0 < x < \infty$). We shall now address the $x \to 0$ limit and the $x \to \infty$ limit of the log-Gibbs density.

Set

$$\epsilon_+ = -\frac{1}{\tau} G(-\infty) = -\frac{a_-}{\tau}, \tag{17.32}$$

and consider the $x \to 0$ limit:

$$\lim_{x \to 0} \frac{c_* \exp\left\{-\frac{1}{\tau} V[\ln(x)]\right\} \cdot \frac{1}{x}}{x^{\epsilon_+ - 1}} \tag{17.33}$$

(using the change-of-variables $x \mapsto z = \ln(x)$)

$$= \lim_{z \to -\infty} \frac{c_* \exp\left[-\frac{1}{\tau} V(z)\right]}{\exp(\epsilon_+ z)} \tag{17.34}$$

(using some algebra and Eq. (17.32))

$$= c_* \exp\left\{-\frac{1}{\tau} \lim_{z \to -\infty} [V(z) - a_- z]\right\} \tag{17.35}$$

(using Eq. (17.27))

$$= c_* \exp\left\{-\frac{1}{\tau} b_-\right\}. \tag{17.36}$$

Equations (17.33)–(17.36) imply that, in the limit $x \to 0$, the log-Gibbs density is asymptotically equivalent to the power function $x^{\epsilon_+ - 1}$ ($x > 0$); this proves the $x \to 0$ part of Eqs. (17.14)–(17.15).

Set

$$\epsilon_- = \frac{1}{\tau} G(\infty) = \frac{a_+}{\tau}, \tag{17.37}$$

and consider the $x \to \infty$ limit:

$$\lim_{x \to \infty} \frac{c_* \exp\left\{-\frac{1}{\tau} V[\ln(x)]\right\} \cdot \frac{1}{x}}{x^{-\epsilon_- - 1}} \tag{17.38}$$

(using the change-of-variables $x \mapsto z = \ln(x)$)

$$= \lim_{z \to \infty} \frac{c_* \exp\left[-\frac{1}{\tau} V(z)\right]}{\exp(-\epsilon_- z)} \tag{17.39}$$

(using some algebra and Eq. (17.37))

$$= c_* \exp\left\{-\frac{1}{\tau} \lim_{z \to \infty} [V(z) - a_+ z]\right\} \tag{17.40}$$

(using Eq. (17.28))

$$= c_* \exp\left\{-\frac{1}{\tau} b_+\right\}. \tag{17.41}$$

Equations (17.38)–(17.41) imply that, in the limit $x \to \infty$, the log-Gibbs density is asymptotically equivalent to the power function $x^{-\epsilon_- - 1}$ ($x > 0$); this proves the $x \to \infty$ part of Eqs. (17.14)–(17.15).

References

1. T. Piketty, *Le Capital au XXIe siecle* (Editions du Seuil, Paris, 2013); *Capital in the Twenty-First Century* (Harvard University Press, Boston, 2014)
2. J.F.C. Kingman, *Poisson Processes* (Oxford University Press, Oxford, 1993)
3. W.J. Reed, Econ. Lett. **74**, 15 (2001)
4. W.J. Reed, B.D. Hughes, Phys. Rev. E **66**, 067103 (2002)
5. W.J. Reed, Phys. A **319**, 469 (2003)
6. W.J. Reed, M. Jorgensen, Commun. Stat. Theory Methods **33**, 1733 (2004)
7. I. Eliazar, M.H. Cohen, Phys. A **391**, 5598 (2012)
8. I. Eliazar, M.H. Cohen, J. Phys. A: Math. Theor. **46**, 365001 (2013)
9. C.P. Stark, N. Hovius, Geophys. Res. Lett. **28**, 1091 (2001)
10. W.J. Reed, J. Reg. Sci. **42**, 1 (2002)

References

11. W. Li, X. Cai, Phys. Rev. E **69**, 046106 (2004)
12. W.J. Reed, in *Advances in Distribution Theory, Order Statistics, and Inference*, eds. by N. Balakrishnan, E. Castillo, J.M. Sarabia (Birkhauser, Boston, 2006), pp. 61–74
13. M. Guida, F. Maria, Chaos, Solitons Fractals **31**, 527 (2007)
14. K. Giesen, A. Zimmermann, J. Suedekum, J. Urban Econ. **68**, 129 (2010)
15. B. Ribeiro, W. Gauvin, B. Liu, D. Towsley, in *INFOCOM IEEE Conference on Computer Communications Workshops*, IEEE (2010), p. 1
16. A.A. Toda, Phys. Rev. E **83**, 046122 (2011)
17. A.A. Toda, J. Econ. Behav. Org. **84**, 364 (2012)
18. Z. Fang, J. Wang, B. Liu, W. Gong, in *Handbook of optimization in complex networks*, eds. by M.T. Thai, P.M. Pardalos (Springer, Boston, 2012), pp. 55–80
19. G. Hajargasht, W.E. Griffiths, Econ. Model. **33**, 593 (2013)
20. C. Wang, X. Guan, T. Qin, T. Yang, Inform. Sci. **330**, 186 (2016)
21. A.A. Toda, Macroecon. Dyn. **21**, 1508 (2017)
22. J. Luckstead, S. Devadoss, Phys. A **465**, 573 (2017)
23. I. Eliazar, M.H. Cohen, Phys. Rev. E **88**, 052104 (2013)
24. I. Eliazar, Phys. A **492**, 123 (2018)
25. A. Friedman, *Stochastic Differential Equations and Applications* (Dover, New York, 2006)
26. B. Oksendal, *Stochastic Differential Equations: An Introduction with Applications*, 6th edn. (Springer, New York, 2010)
27. L. Arnold, *Stochastic Differential Equations: Theory and Applications* (Dover, New York, 2011)
28. W.T. Coffey, Yu.P Kalmykov, J.T. Waldron, *The Langevin Equation* (World Scientific, Singapore, 2012)
29. P. Langevin, Comptes Rendus Acad. Sci. (Paris) **146**, 530 (1908)
30. C. Gardiner, *Handbook of Stochastic Methods* (Springer, New York, 2004)
31. N.G. Van Kampen, *Stochastic Processes in Physics and Chemistry*, 3rd edn. (North-Holland, Amsterdam, 2007)
32. H.O. Georgii, *Gibbs Measures and Phase Transitions* (De Gruyter, Berlin, 2011)
33. H. Risken, *The Fokker-Planck Equation: Methods of Solutions and Applications* (Springer, New York, 1996)
34. E.T. Jaynes, Phys. Rev. **106**, 620 (1957)
35. E.T. Jaynes, Phys. Rev. **108**, 171 (1957)
36. N. Wu, *The Maximum Entropy Method* (Springer, New York, 1997)
37. J.N. Kapur, *Maximum-Entropy Models in Science and Engineering* (New Age, New Delhi, 2009)
38. K. Ito, H.P. McKean, *Diffusion processes and their sample paths*, Reprint of the 1974th edn. (Springer, Berlin, 1996)
39. I. Karatzas, S. Shreve, *Brownian Motion and Stochastic Calculus* (Springer, New York, 1991)
40. J.F. Le-Gall, *Brownian Motion, Martingales, and Stochastic Calculus* (Springer, New York, 2016)
41. I. Eliazar, Phys. Rep. **649**, 1 (2016)
42. I. Eliazar, Ann. Phys. **389**, 306 (2018)
43. J. Aitchison, J.A.C. Brown, *The Lognormal Distribution with Special Reference to its Uses in Econometrics* (Cambridge University Press, Cambridge, 1957)
44. E.L. Crow, K. Shimizu (eds.), *Lognormal Distributions: Theory and Applications* (Marcel Dekker, New York, 1988)
45. J.K. Patel, C.B. Read, *Handbook of the Normal Distribution* (Dekker, New York, 1996)
46. S. Kotz, T. Kozubowski, K. Podgorski, *The Laplace Distribution and Generalizations* (Springer, New York, 2012)

Chapter 18
Conclusion

In this monograph we presented a comprehensive in-depth study of the power Poisson processes $\mathcal{E}_+$ and $\mathcal{E}_-$. These Poisson processes are defined over the positive half-line $(0, \infty)$, and are governed by power intensity functions. Specifically: the power Poisson process $\mathcal{E}_+$ is characterized by the intensity function $\lambda_+(x) = c\epsilon x^{\epsilon-1}$ ($x > 0$), and the power Poisson process $\mathcal{E}_-$ is characterized by the intensity function $\lambda_-(x) = c\epsilon x^{-\epsilon-1}$ ($x > 0$); the coefficient c and the exponent ϵ of these intensity functions are positive parameters.

The power Poisson process $\mathcal{E}_-$ has finitely many points below any positive level l, and infinitely many points above. Consequently, the points of the power Poisson process $\mathcal{E}_+$ can be represented as a sequence of monotone increasing order statistics that diverge to infinity: $P_+(1) < P_+(2) < P_+(3) < \cdots \nearrow \infty$. Conversely, the power Poisson process $\mathcal{E}_-$ has finitely many points above any positive level l, and infinitely many points below. Consequently, the points of the power Poisson process $\mathcal{E}_-$ can be represented as a sequence of monotone decreasing order statistics that converge to zero: $P_-(1) > P_-(2) > P_-(3) > \cdots \searrow 0$.

Limit laws presented in Chaps. 2 and 13 established the emergence of the power Poisson processes $\mathcal{E}_+$ and $\mathcal{E}_-$. Detailed analyses presented in the other chapters explored the power Poisson processes $\mathcal{E}_+$ and $\mathcal{E}_-$ from diverse perspectives: structural, statistical, fractal, dynamical, socioeconomic, and stochastic. In particular, these analyses affirmed what was proclaimed in the introduction: **The power Poisson processes $\mathcal{E}_+$ and $\mathcal{E}_-$ are the foundational "bedrock" over which an assortment of power statistics stand**.

As figuratively described in the introduction: the power Poisson processes $\mathcal{E}_+$ and $\mathcal{E}_-$ are infinite-dimensional "icebergs", and finite-dimensional "tips" of these icebergs from an assortment of power statistics. Here is a terse summary of the principal power statistics encountered along the monograph.

Power Poisson law. Consider the number of points $N_+(l)$ of the power Poisson process $\mathcal{E}_+$ that reside below the positive level l. The number $N_+(l)$ is a Poisson random variable that is characterized by the mean

$$cl^\epsilon .$$

Power Poisson law. Consider the number of points $N_-(l)$ of the power Poisson process $\mathcal{E}_-$ that reside above the positive level l. The number $N_-(l)$ is a Poisson random variable that is characterized by the mean

$$cl^{-\epsilon} .$$

Log-log plot. Consider the number of points $N_+(l)$ of the power Poisson process $\mathcal{E}_+$ that reside below the positive level l. Further consider a log-log plot of the log-number $\ln[N_+(l)]$ versus the log-level $\ln(l)$ ($l > 0$). This log-log plot is approximated by the affine line

$$y = \ln(c) + \epsilon x .$$

Log-log plot. Consider the number of points $N_-(l)$ of the power Poisson process $\mathcal{E}_-$ that reside above the positive level l. Further consider a log-log plot of the log-number $\ln[N_-(l)]$ versus the log-level $\ln(l)$ ($l > 0$). This log-log plot is approximated by the affine line

$$y = \ln(c) - \epsilon x .$$

Inverse-Pareto law. Sample at random a point from the (finitely many) points of the power Poisson process $\mathcal{E}_+$ that reside below the positive level l. The position $B_+(l)$ of the randomly sampled point is an inverse-Pareto random variable that is characterized by the cumulative distribution function

$$(x/l)^\epsilon$$

($0 < x < l$).

Pareto law. Sample at random a point from the (finitely many) points of the power Poisson process $\mathcal{E}_-$ that reside above the positive level l. The position $A_-(l)$ of the randomly sampled point is a Pareto random variable that is characterized by the tail distribution function

$$(l/x)^\epsilon$$

($l < x < \infty$).

Weibull law. Consider the position $A_+(0)$ of the smallest point (i.e., the minimal point) of the power Poisson process $\mathcal{E}_+$. The position $A_+(0)$ is a Weibull random variable that is characterized by the forward hazard function

$$c\epsilon x^{\epsilon-1}$$

18 Conclusion

($x > 0$).

Fréchet law. Consider the position $B_-(\infty)$ of the largest point (i.e., the maximal point) of the power Poisson process $\mathcal{E}_-$. The position $B_-(\infty)$ is a Fréchet (inverse-Weibull) random variable that is characterized by the backward hazard function

$$c\epsilon x^{-\epsilon-1}$$

($x > 0$).

Power order statistics. Consider the position $P_+(k)$ of the kth order statistic of the power Poisson process $\mathcal{E}_+$. The mean and the mode of the random variable $P_+(k)$ are both asymptotically equivalent to

$$(k/c)^{1/\epsilon}$$

(as $k \to \infty$).

Power order statistics. Consider the position $P_-(k)$ of the kth order statistic of the power Poisson process $\mathcal{E}_-$. The mean and the mode of the random variable $P_-(k)$ are both asymptotically equivalent to

$$(c/k)^{1/\epsilon}$$

(as $k \to \infty$).

Inverse-Zipf law. Consider the position $P_+(k)$ of the kth order statistic of the power Poisson process $\mathcal{E}_+$. Further consider a log-log plot of the log-position $\ln[P_+(k)]$ versus the log-rank $\ln(k)$ ($k = 1, 2, 3, \ldots$). This log-log plot is approximated by the affine line

$$y = \frac{1}{\epsilon}[x - \ln(c)] .$$

Zipf law. Consider the position $P_-(k)$ of the kth order statistic of the power Poisson process $\mathcal{E}_-$. Further consider a log-log plot of the log-position $\ln[P_-(k)]$ versus the log-rank $\ln(k)$ ($k = 1, 2, 3, \ldots$). This log-log plot is approximated by the affine line

$$y = \frac{1}{\epsilon}[\ln(c) - x] .$$

Zipf law. Consider the kth consecutive ratio $P_+(k+1)/P_+(k)$ of the order statistics of the power Poisson process $\mathcal{E}_+$, as well as the kth consecutive ratio $P_-(k)/P_-(k+1)$ of the order statistics of the power Poisson process $\mathcal{E}_-$. Further consider a log-log plot of the log-ratio $\ln[P_+(k+1)/P_+(k)]$ versus the log-rank $\ln(k)$ ($k = 1, 2, 3, \ldots$), and the log-log plot of the log-ratio $\ln[P_-(k)/P_-(k+1)]$ versus the log-rank $\ln(k)$ ($k = 1, 2, 3, \ldots$). Both these log-log plots are approximated by the affine line

$$y = -\ln(\epsilon) - x .$$

Power Lorenz curve. Consider the points of the power Poisson process $\mathcal{E}_+$ to represent the wealth values of the members of a "virtual" society. The proportion of the society's wealth that is held by the low (poor) $100u\%$ of the society's members is governed by the Lorenz curve

$$u^{1+1/\epsilon}$$

$(0 \le u \le 1)$.

Power Lorenz curve. Consider the points of the power Poisson process $\mathcal{E}_-$ to represent the wealth values of the members of a "virtual" society. The proportion of the society's wealth that is held by the top (rich) $100u\%$ of the society's members is governed by the Lorenz curve

$$u^{1-1/\epsilon}$$

$(0 \le u \le 1)$; in this Lorenz curve the exponent ϵ is restricted to the range $\epsilon > 1$.

One-sided Lévy law. Consider the sum Σ_1 of the points of the power Poisson process $\mathcal{E}_-$. The sum Σ_1 is a one-sided Lévy random variable that is characterized by the log-Laplace transform

$$c_1 \theta^\epsilon$$

$(\theta \ge 0)$, where c_1 is a positive coefficient; in this log-Laplace transform the exponent ϵ is restricted to the range $\epsilon < 1$.

Symmetric Lévy law. Consider the sum Σ_2 of the points of the power Poisson process $\mathcal{E}_-$, after having attached random signs to these points. The sum Σ_2 is a symmetric Lévy random variable that is characterized by the log-Fourier transform

$$c_2 |\theta|^\epsilon$$

$(-\infty < \theta < \infty)$, where c_2 is a positive coefficient; in this log-Fourier transform the exponent ϵ is restricted to the range $\epsilon < 2$.

Power Newcomb–Benford law. Consider the first digit, in a power expansion with base b, of the points of the power Poisson process $\mathcal{E}_+$. The probability that the first digit is smaller than the integer i is

$$\frac{i^\epsilon - 1}{b^\epsilon - 1}$$

$(i = 1, \ldots, b)$.

Power Newcomb–Benford law. Consider the first digit, in a power expansion with base b, of the points of the power Poisson process $\mathcal{E}_-$. The probability that the first digit is smaller than the integer i is

$$\frac{1 - i^{-\epsilon}}{1 - b^{-\epsilon}}$$

$(i = 1, \ldots, b)$.

18 Conclusion

Sub-diffusion and super-diffusion. Consider the aggregate of IID random motions (real-valued and with zero mean), where the motions' inherent time-paces are the points of the power Poisson process $\mathcal{E}_-$. The temporal mean-square-displacement of the aggregate motion is

$$c_* t^\epsilon$$

($t \geq 0$), where c_* is a positive coefficient.

Flicker noise and 1/f noise. Consider the aggregate of IID stationary velocity processes (real-valued and with zero mean), where the processes' inherent time-paces are the points of the power Poisson process $\mathcal{E}_-$. The spectral density of the aggregate process is

$$c_* |\theta|^{1-\epsilon}$$

($-\infty < \theta < \infty$), where c_* is a positive coefficient; in this spectral density the exponent ϵ is restricted to the range $\epsilon > 1$.

Herdan–Heaps law. Consider the first-passage-times (FPTs) of IID random motions in a general space, where the motions' inherent time-paces are the points of the power Poisson process $\mathcal{E}_-$. These FPTs form a power Poisson process $\mathcal{E}_+$ with temporal innovation rate

$$c_* t^\epsilon$$

($t \geq 0$), where c_* is a positive coefficient.

Each of the above equations manifests a different type of a statistical power-law relation. A priori, this assortment of power statistics—which are observed across science and engineering—looks like a "zoo" of statistical power-law forms. A posteriori, we see that all these power statistics are founded on common bedrock: the power Poisson processes $\mathcal{E}_+$ and $\mathcal{E}_-$.

Power statistics are ubiquitous and are of profound importance. This monograph provides an overarching panoramic exposition of power statistics. This exposition is based on a unified "bedrock approach": a comprehensive exploration of the power Poisson processes $\mathcal{E}_+$ and $\mathcal{E}_-$. With its scope and depth, this monograph is poised to serve researchers and practitioners—from various fields of science and engineering—that are engaged in power-statistics analyses.